Speech in Mobile and Pervasive Environments

Wiley Series on Wireless Communications and Mobile Computing

Series Editors: Dr Xuemin (Sherman) Shen, *University of Waterloo, Canada*
Dr Yi Pan, *Georgia State University, USA*

The "Wiley Series on Wireless Communications and Mobile Computing" is a series of comprehensive, practical and timely books on wireless communication and network systems. The series focuses on topics ranging from wireless communication and coding theory to wireless applications and pervasive computing. The books provide engineers and other technical professionals, researchers, educators, and advanced students in these fields with invaluable insight into the latest developments and cutting-edge research.

Other titles in the series:

Misic and Misic: Wireless Personal Area Networks: Performance, Interconnection, and Security with IEEE 802.15.4, *January 2008, 978-0-470-51847-2*

Takagi and Walke: Spectrum Requirement Planning in Wireless Communications: Model and Methodology for IMT-Advanced, *April 2008, 978-0-470-98647-9*

Pérez-Fontán and Espiñeira: Modeling the Wireless Propagation Channel: A simulation approach with MATLAB®, *August 2008, 978-0-470-72785-0*

Ippolito: *Satellite Communications Systems Engineering: Atmospheric Effects, Satellite Link Design and System Performance*, August 2008, 978-0-470-72527-6

Lin and Sou: *Charging for Mobile All-IP Telecommunications*, September 2008, 978-0-470-77565-3

Myung and Goodman: *Single Carrier FDMA: A New Air Interface for Long Term Evolution*, October 2008, 978-0-470-72449-1

Wang, Kondi, Luthra and Ci: *4G Wireless Video Communications*, April 2009, 978-0-470-77307-9

Cai, Shen and Mark: *Multimedia Services in Wireless Internet: Modeling and Analysis*, June 2009, 978-0-470-77065-8

Stojmenovic: *Wireless Sensor and Actuator Networks: Algorithms and Protocols for Scalable Coordination and Data Communication*, February 2010, 978-0-470-17082-3

Liu and Weiss, *Wideband Beamforming: Concepts and Techniques*, March 2010, 978-0-470-71392-1

Riccharia and Westbrook, *Satellite Systems for Personal Applications: Concepts and Technology*, July 2010, 978-0-470-71428-7

Qian, Muller and Chen: *Security in Wireless Networks and Systems*, March 2014, 978-0-470-512128

Speech in Mobile and Pervasive Environments

Nitendra Rajput and Amit A. Nanavati

IBM Research, New Delhi, India

A John Wiley & Sons, Ltd., Publication

This edition first published 2012

Registered office
John Wiley & Sons Ltd, The Atrium, Southern Gate, Chichester, West Sussex, PO19 8SQ, United Kingdom

For details of our global editorial offices, for customer services and for information about how to apply for permission to reuse the copyright material in this book please see our website at www.wiley.com.

Library of Congress Cataloging-in-Publication Data

Rajput, Nitendra.
Speech in mobile and pervasive environments / Nitendra Rajput and Amit A. Nanavati.
p. cm.
Includes bibliographical references and index.
ISBN 978-0-470-69435-0 (cloth)
1. Speech processing systems. 2. Cell phone systems. I. Nanavati, Amit A. II. Title.
TK7882.S65R334 2012
006.5–dc23

2011033626

A catalogue record for this book is available from the British Library.

ISBN: 9780470694350 (H/B)

Typeset in 10.5/13pt Times by Laserwords Private Limited, Chennai, India

FOR,

Zo͡ozo͡o & Po⊤o

TO,

Family & Friends

Contents

About the Series Editors

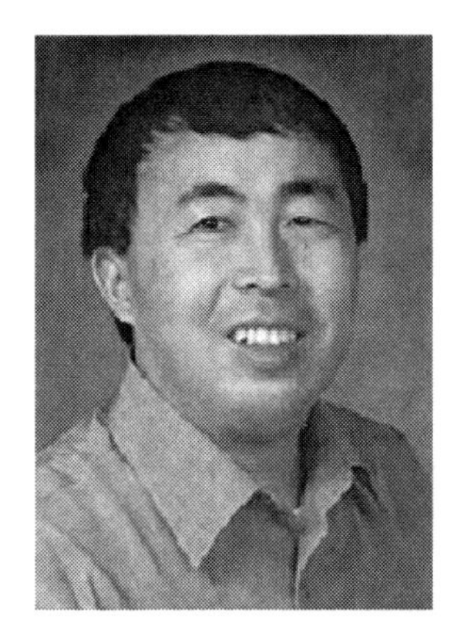

Xuemin (Sherman) Shen (M'97-SM'02) received a BSc degree in electrical engineering from Dalian Maritime University, China in 1982, and the MSc and PhD degrees (both in electrical engineering) from Rutgers University, New Jersey, USA, in 1987 and 1990 respectively. He is a Professor and University Research Chair, and the Associate Chair for Graduate Studies, at the Department of Electrical and Computer Engineering, University of Waterloo, Canada. His research focuses on mobility and resource management in interconnected wireless/wired networks, UWB wireless communications systems, wireless security, and ad hoc and sensor networks. He is a co-author of three books, and has published more than 300 papers and book chapters on wireless communications and networks, control and filtering. Dr. Shen serves as a founding area editor for *IEEE Transactions on Wireless Communications*; editor-in-chief for *Peer-to-Peer Networking and Application*; associate editor for *IEEE Transactions on Vehicular Technology, KICS/IEEE Journal of Communications and Networks, Computer Networks, ACM/Wireless Networks* and *Wireless Communications and Mobile Computing*. He has also served as a guest editor for *IEEE JSAC, IEEE Wireless Communications* and *IEEE Communications Magazine*. Dr. Shen received the Excellent Graduate Supervision Award in 2006, and the Outstanding Performance Award in 2004 from the University of Waterloo, the Premier's Research Excellence Award (PREA) in 2003 from the Province of Ontario, Canada, and the Distinguished Performance Award in 2002 from the Faculty of Engineering, University of Waterloo. Dr. Shen is a registered Professional Engineer of Ontario, Canada.

Yi Pan is the Chair and a Professor in the Department of Computer Science at Georgia State University, USA. Dr Pan received his BEng and MEng degrees in computer engineering from Tsinghua University, China, in 1982 and 1984, respectively, and his PhD degree in computer science from the University of Pittsburgh, USA, in 1991. Dr. Pan's research interests include parallel and distributed computing, optical networks, wireless networks and bioinformatics. Dr. Pan has published more than 100 journal papers, with over 30 papers published in IEEE journals. In addition, he has published over 130 papers in refereed conferences (including IPDPS, ICPP, ICDCS, INFOCOM, and GLOBECOM). He has also co-edited over 30 books. Dr. Pan has served as an editor-in-chief or an editorial board member for 15 journals including five IEEE Transactions journals and has organized many international conferences and workshops. Dr. Pan has delivered over 10 keynote speeches at international conferences. Dr. Pan is an IEEE Distinguished Speaker (2000–2002), a Yamacraw Distinguished Speaker (2002), and a Shell Oil Colloquium Speaker (2002). He is listed in *Men of Achievement, Who's Who in America, Who's Who in American Education, Who's Who in Computational Science and Engineering*, and *Who's Who of Asian Americans*.

List of Contributors

Patrick J. Bourke, Carnegie Mellon University, USA

Om D. Deshmukh, IBM Research, India

Jaakko Hakulinen, University of Tampere, Finland

Amit A. Nanavati, IBM Research, India

Nitendra Rajput, IBM Research, India

Rob A. Rutenbar, Carnegie Mellon University, USA

Markku Turunen, University of Tampere, Finland

Kai Yu, Carnegie Mellon University, USA

Foreword

Advances in computing–in terms of both the creation of novel mathematical techniques and the design of data-driven technologies–have fuelled the ubiquitous development and deployment of speech technologies over the last two decades. Some of the core speech technologies and their applications to coding, recognition, synthesis, enhancement and such have been well documented in several excellent books, and have been since incorporated in university course across the world. With the recent trends toward convergence of computing and communication, well exemplified by the global proliferation of mobile devices in the past decade, there has been significant speech technology research and development effort focused on algorithms and multimodal interfaces applications targeting and tailored to the new requirements of mobile platforms and interaction paradigms. This book is a natural and timely contribution that summarizes the state of the art in this domain of mobile speech technologies that can be useful both as teaching resource as well as a reference to the practitioner.

In this book, the authors have done a fantastic job in providing a comprehensive account, offering an end-to-end story for enabling speech interfaces on mobile devices. The challenge in undertaking to write a book of this nature is to have the right breadth to do justice to a multifaceted area. As the founders and organizers of the yearly workshops on Speech in Mobile and Pervasive Environments since 2006, the authors were indeed well positioned to take on this challenging assignment. I have been a close and participating witness to the growth of interest and the forming of a community in this field. Furthermore, I believe that that several core technology pieces have matured well to warrant the deployment of speech applications on mobile devices in the marketplace. It is hence I AM delighted to see a book on this topic at this time.

While on the one hand mobile and pervasive environments provide an opportunity to build novel speech-based applications but on the other, the computing limitations of the device pose a challenge to the design and implementation of the underlying speech technology. This book provides an excellent overview of both the technologies from a device point of view and the possible applications from an enabling technology point of view. Chapters 2, 3 and 4 describe the implications and research in core speech technology owing to the mobile device restrictions. Chapters 5 and 6 move towards the application side of the story and describe the special requirements of designing speech applications on mobile devices. Chapters 7, 8 and 9 focus on the possible speech based applications on mobile devices. The earlier chapters are structured such that there is a natural transition from the descriptions of core technology that refer to the standard ways of speech recognition and synthesis to those tailored for the mobile device. Similarly, the later chapters on applications constantly refer to specific real world solutions to drive home the vast technology possibilities.

I see this book to have a wide audience in both academia and industry. In the university, it can be used for advanced courses on mobile technologies and speech technologies, targeting not just engineers but also application designers and mobile technology developers. Students can pick a specific sub-area (either hardware or speech recognition or speech synthesis or interface design) and do course projects. I also believe this book to be of value to industry since it brings together both the core technology and the wide range of application possibilities.

I hope that the reader will be able to appreciate and benefit from the comprehensive account that aims to bridge technology and applications. This book does well to achieve the dual goals as an educational resource for the student and as a reference for the practitioner. This book is indeed a compelling and useful contribution to the speech research and development world.

Shrikanth Narayanan
University of Southern California

Preface

What this book is about

If early 21st century is to be remembered for one global phenomenon, the rise of the mobile phone is a good contender. Mobiles are here to stay. *Speech in Mobile and Pervasive Environments* deals with issues related to speech processing on resource-constrained, wireless, mobile devices including: speech recognition in noisy environments, specialized hardware for speech recognition and synthesis, the use of context to enhance recognition, the emerging new standards required for interoperability, multimodal applications on mobile devices, distributed processing between the client and the server, and the relevance of speech for developing regions.

While speech processing has been an independent area of research for many years, the proliferation of the mobile device is making speech mainstream. Many novel and interesting applications are being offered on mobile devices, and the mobile platform brings its own opportunities and constraints: new sources of context, resource constraints and noisy environments. There is a rapidly growing interest, as indicated in academic conferences, as well as considerable investment in the telecoms and related industries.

Why we wrote this book

A multi-disciplinary approach to this topic is the primary motivation for writing this book. Further,

- The mobile is a convergent platform – a personal communication device, and an interface for applications and services.

- We feel that it is useful for speech recognition experts to be able to leverage context and for people working in developing regions to learn about embedded recognition; it is therefore helpful to have all these topics between two covers.

- The topics covered in this book vary widely in flavor (from distributed speech recognition to applications in developing regions) and in maturity (the oldest being hardware and developing regions being the youngest). As a result of this, the chapters are not uniform in length or presentation, or in the approach we have taken toward the selection of material.
- Even for topics that are well-established, we hope that the expert will find something of use in this book. For example, rather than cover the issues that generally arise in gathering and managing context, our focus here has been to address context as it relates to speech in mobile and pervasive environments.
- We hope that this book will fulfill its ambition of bringing together people working across disciplines, increasing interactions among them and advancing this field further.

Chapters

The chapters of the book are as follows.

- The chapter on hardware looks in detail at speech recognition from a hardware-centric viewpoint. It investigates the possibility of rendering the recognizer in the hardware itself.
- The chapter on embedded recognition and synthesis provides an overview and discusses acoustic parameterization, speech recognition algorithms and acoustic modeling.
- The chapter on distributed speech recognition includes the various protocols and standards used for distributing the recognition effort.
- The chapter on context discusses ways of modeling context and aggregating it for use in speech applications. It also describes a few context-based applications.
- The chapter on software talks about infrastructure, standards and technologies. VoIP and VoiceXML are a part of this chapter. Some possible extensions to VoiceXML from a mobile-speech perspective are also explored. This chapter also considers the question of restructuring a call-flow so that it can be accommodated on devices with constraints.

- The chapter on multimodal dialogue systems provides some examples of distributed multimodal applications and the software architectures needed to support them.
- The chapter on evaluation describes the need to develop appropriate methods for evaluating mobile speech systems. Generic methods of usability studies are discussed first, followed by a consideration of more specific subjects relevant to the evaluation of mobile and speech-based systems. A theoretical measure for dialogue call-flow is also presented.
- A large number of people on our planet cannot read or write, but find mobile telephones very usable and useful. The chapter on developing regions focusses on the needs of this population and the applications and interfaces that are needed to serve them.

The audience for this book

Students

This book can be used as a textbook for a one-semester postgraduate or advanced undergraduate course on the subject for computer science, electrical engineering, and human–computer-interface students. The book might be used as follows.

- A one-semester course for electrical engineers could include the chapters on hardware (Chapter 2), embedded recognition and synthesis (Chapter 3), distributed speech recognition (Chapter 4), multimodality (Chapter 7) and developing regions (Chapter 9). While the first three chapters are core, the last two could feed back into the design of hardware.
- A one-semester course for computer science students could include the chapters on distributed speech recognition (Chapter 4), software (Chapter 6), multimodality (Chapter 7), context in conversation (Chapter 5) and developing regions (Chapter 9).
- A one-semester course for human–computer-interface students could include context in conversation (Chapter 5), multimodality (Chapter 7), software (Chapter 6), evaluation (Chapter 8) and developing regions (Chapter 9).

Professionals

There is a large number of professionals, across many disciplines, who are creating applications and solutions for the mobile platform. Many domain-specific (e.g. healthcare, agriculture) mobile application developers and software developers in the mobile/telecoms industries will also find this book useful.

Nitendra Rajput
Amit A. Nanavati

Acknowledgments

We are extremely grateful to our collaborators, because of whom we had enough to say: Rob, Patrick and Kai of Carnegie Mellon University; We still recall the enthusiastic response we received from Rob for the book, and how he thought this effort was very timely. We deeply appreciate their following through in such a timely manner. To the ever-supportive Marrku and Jakko, who quickly suggested their topics when we talked to them about our proposal, and were prompt with the drafts. Also to Om, who became our saviour in the hour of need – we wonder where he hides his white wings. Our deepest and more heartfelt gratitude to each one of you! When you decide to write your books, think of us:-)

The wisdom that writing a book is not at all like what one thinks it is going to be before writing it, came to us too. Enabling you to read these words took not just patience and drive but also dealing with promotions, transfers and happy family additions on the part of many of our friends at Wiley: Sarah Tilley, Susan Barclay and Sarah Hinton, who helped us conclude that patience is indeed a virtue (we are still alive), and Anna Smart, who risked her career and braved a couple of engineers' suggestions on cover design. We are grateful to our friends at Laserwords and Anglosphere Editing for their meticulous efforts.

Nitendra: To Samvit, who timed his arrival to perfection to ensure that I would have yet another excuse to give my coauthor and the publishers. To Praneeta, who has been patient throughout my life – and with respect to book as well – and never asked tough questions regarding the snail's pace of my writing. To my parents, who always maintain their faith in me in whatever I undertake – a faith that I have not yet been able to extend to my son.

Amit: To my uncle, Prof. H.C. Dholakia, who wrote ten odd (well, law) law books, and misled me into feeling that it was easy to write one. To my young nephews Arnav and Setu, who, through their innocent questioning, made us realize that my uncle had actually cheated us. Poto, for gently advising and encouraging me to complete this work – it would be nice if he took his own advice more often. And most of all, my *Teen Deviyaan*, who define me.

Also to our collaborators in our various papers that lead us into this area and finally to this effort. To our fathers-in-law who were unblushingly punctual in checking on the progress. Finally, to the umpteen friends and all our family members who will start showering their wishes on us as soon as they see this in print – without ever worrying about its content. In the end it is only for these wishes that we have toiled so hard, truly, in embarking upon this adventure.

Nitendra Rajput and Amit A. Nanavati

1

Introduction

Nitendra Rajput and Amit A. Nanavati
IBM Research, India

Speech is the most natural and widely used modality of interaction. In the devices world, mobile phones have surpassed television by a huge margin. Mobile devices and speech interaction therefore form a uniquely pervasive platform to access any information technology application. This book addresses the technology related to speech interaction on such mobile and pervasive devices. While speech interfaces and mobile devices have separately been a key areas of research and study, the two together have not been studied together in such detail. This is the gap that the book wishes to address.

Speech in mobile and pervasive environments is an exciting and a very promising area from the perspectives of timing, technology and adoption. We will now elaborate these three perspectives.

Increased processing speeds and reduced dimensions of computing devices over the last two decades have made computing more pervasive. While the first computer ever made (ENIAC in the 1940s) was of the size of a warehouse, the same amount of computing is now possible in chips that are measured in millimeters. The effect of Moore's law in reducing the size and increasing computational speed is clearly evident in the computing devices that are available today. The computers in 1950s were mostly huge machines used by enterprises or research organizations. By the 1990s, most computers in the world were personal computers that people used in home or offices. The last decade has

Speech in Mobile and Pervasive Environments, First Edition.
Nitendra Rajput and Amit A. Nanavati.

seen the computing world move to laptops and smart phones. This journey of reduced size and increased speed in the computing world has been complemented by a corresponding acceptance of more personalized and pervasive devices in the consumer space. Computing has thus penetrated every aspect of human life. Devices that are easy to use and carry have increased the pervasiveness of technology. *The timing is therefore right to look at such devices, which are the doors enabling access to the computing world in the 2010s*.

From a technology perspective, research in speech technologies started its journey way back in the 1950s in the area of digit recognition[1] and later through the IBM Shoebox. The technology has advanced significantly since then and now speech recognition and synthesis are becoming increasingly available on mobile devices. Speech recognition systems now understand natural language and use the context of the application to increase the recognition accuracy. Speech recognition and synthesis are also available in a large number of languages. *The technology has now reached such a level of maturity that we can consider speech as a main medium of interaction on the mobile platform*.

In terms of adoption, mobile devices clearly outnumber any other computing devices. The phenomenon is not only true in the developed nations, but also in developing regions, where cost and literacy are a challenge. We spend a significant amount of time on the move. Smart phones and other pervasive computing technologies are finding their way in the market to address the needs of such people who are not always near a computer. Not only do these devices provide comfortable access, but they also penetrate the market to reach out to populations who do not necessarily use computers. Pervasive environments, including mobile devices, are therefore now connecting more people on this planet than computers did. Being pervasive, such devices are more accessible than their counterpart computers. So not only do more users now own such devices, they are accessing them for longer durations than they would on traditional computers.

In such an evolving environment, *mobile and pervasive devices are expected to attract attention of most users for maximum time*.

We tend to rely on a keyboard to interact with computers and mobile devices. Keyboards are designed from a machine perspective and the machine is able to parse any keyboard entry with complete accuracy. However keyboards are not a natural means of interaction for humans. Other natural modalities such as gestures and emotions are easier for a human, but parsing such inputs is difficult for a machine, given the state of the art. *Speech provides a good balance, being natural to humans and still parsable by the machine*.

[1] Davies, K.H., Biddulph, R. and Balashek, S. (1952) Automatic Speech Recognition of Spoken Digits, J. Acoust. Soc. Am. 24(6), pp. 637–642.

> Using speech as a medium of interaction with the pervasive environment therefore promises to be a very natural and highly usable environment for humans to interact with machines.

From a technical perspective, speech on mobile devices has orthogonal implications on the entire life-cycle of the system. This includes the application design, the modality of interaction, the processing of interaction and the eventual evaluation. We will now look at the implications on each of these components of the life-cycle and present the manner in which they need to be addressed from a speech and a mobile perspective. These discussions will enable a reader to relate to each of these topics and then relate to the chapters of the book that describe them. The goal of this chapter is therefore twofold: we want to introduce the reader to the challenges associated with speech in mobile and pervasive environments and secondly, we describe the layout of the book so that the chapters can be read in a different sequence if preferred.

1.1 Application design

When a mobile application uses the speech modality, it can be designed to derive more intelligence from the context of the user since such devices are mostly personal. The rich context of the user and their environment can make the application communicate more intelligently with the user. An application design should therefore incorporate the context of the user and the environment. Speech interfaces tend to gain further from this context since the machine is able to parse human speech more effectively if it knows the context. The implications of the availability of context and its usage is described in Chapter 5.

Since mobile applications need to run on a variety of devices, it is pertinent that the speech application runs over the multitude of operating systems and that they use standard authoring languages. Chapter 6 focuses on the various standards that have implications for speech applications on mobile devices. It also provides details about the way such standards need to be engineered to enable efficient processing of speech signals on resource-constrained mobile devices.

1.2 Interaction modality

Speech, though natural, is not the most efficient modality in delivering every information. Research in the mobile interactions world has realized that a multimodal interface involving both speech and visual interactions can be a better modality than a speech-only or a visual-only interface. Adding speech modality to any application therefore results in a multimodal application. The architecture and framework of building such multimodal applications therefore needs to be studied in a different way compared to traditional application development interfaces such as web application development frameworks. Several mobile-based applications can either perform the processing on the device or push it to the server. Architectural models that can support applications in such distributed processing modes also need to be studied. We describe the multimodal architecture and illustrate this through several multimodal applications that run on mobile devices. This explanation and more detail of the several client-server architectural models form the basis of Chapter 7.

1.3 Speech processing

Speech interaction itself has many broad areas of research. One of the more intensively studied areas relate to converting user speech to text and converting text to speech. Research in these two areas has been one of the key applications in the core fields of digital signal processing and pattern recognition. Researchers have worked extensively on developing systems that can process a variety of human speech pronunciations and determine the underlying text. The difficulty of this problem is attributed to the nuances of the language and its multiple variations in terms of dialects, varying pronunciations and spontaneity of the speaker. Traditional approaches to solving the speech-to-text problem are therefore compute and data intensive as they try to determine a pattern from a sea of possibilities. Similarly, converting text to natural speech is also data intensive, as this involves determining the most appropriate audio sample from a large sea of possible sounds in a particular language. Speech processing in the mobile world is therefore a non-trivial adaptation from the computer world. Several techniques to efficiently process speech on resource-constrained devices are outlined in Chapters 2 and 3.

Owing to the high computing and data requirements, speech to text and text to speech need novel solutions when they are to be performed in a mobile environment. Most mobile devices cannot support huge amounts of processing and they do not have large amounts of processing memory either. From a mobile context, it therefore becomes critical to build speech recognition systems that can work on resource-constrained devices. Chapter 2 provides a brief overview of a simple speech recognition system and then elaborates the various techniques that are used to reduce the footprint. Embedded speech recognition provides methods to perform speech recognition with least degradation in accuracy but on such low-end devices. The chapter also illustrates techniques to convert text into natural speech by using a smaller data set, which can be incorporated in pervasive devices.

Since a large number of applications are in a client-server environment, a server can also be used to perform some speech processing. There are standards that enable such distribution of speech processing across the client and the server devices. We describe this concept of distributing speech processing in Chapter 4. The chapter also provides details on the underlying protocols that are used to communicate the processing between the client and the server. We compare the different distributed processing techniques at the end of the chapter.

1.4 Evaluations

The last leg of a mobile application is its evaluation. Being multimodal in nature, the evaluations of such an interface requires redesigning of standard evaluation techniques that usually work on specific modalities. Since we are focusing on speech applications in pervasive devices, the evaluations need to mimic the real-world conditions in which such applications are expected to be used. In most pervasive environments, a user is surrounded by other activities in addition to the device. Therefore seeking full attention of the user to the application is not always possible. We outline the various evaluation techniques for multimodal applications in Chapter 8.

We describe the distinction between field studies and real-world studies and describe their use in separate applications.

Toward the end of this book, we present a specific case of speech applications as applied in the context of the developing world. This deserves a separate chapter since we believe that owing to the pervasiveness of these devices and their easy modality, the applications are well suited to low-literate users in the developing world. In Chapter 9 we describe the challenges that are currently associated with current technologies for low-literate users and then outline how speech-based mobile applications can bridge this gap.

The book thus covers the various aspects of a mobile application lifecycle when speech processing will be used. We believe that the reader will be able to get a broad understanding of all the inter-related issues and also a detailed understanding of the specific challenges in this domain. The authors also conduct yearly workshops on this title and the reader is encouraged to participate in these[2] and join the community at our wiki.[3]

[2] http://research.ihost.com/SiMPE.

[3] http://simpe.wikispaces.com.

2

Mobile speech hardware: The case for custom silicon

Patrick J. Bourke, Kai Yu and Rob A. Rutenbar
Carnegie Mellon University, USA

Mobile platforms offer limited computational resources – a significant side effect of constraints on their cost, size and batteries. As a consequence, speech recognizers on mobile platforms usually make one of two unavoidable decisions: (1) to reduce the capability of a software-based recognizer hosted on this platform or (2) to move some part of the recognition computation off this platform.

In this chapter we suggest a third alternative: *render the recognizer itself in hardware on the mobile platform*. This is the path taken by graphics applications such as video playback, which are no longer handled in software, even on mobile phones. This chapter describes the hardware-based solution, explains its novel constraints and opportunities, and discusses recent results targeting low-power custom silicon recognizers.

2.1 Introduction

In this chapter we look in detail at speech recognition from a hardware-centric viewpoint. We must acknowledge up front that our agenda is rather atypical

Speech in Mobile and Pervasive Environments, First Edition.
Nitendra Rajput and Amit A. Nanavati.

in this arena. That is, we will not focus on the manifest problems of putting a recognizer *on* the hardware of a specific mobile platform, but instead focus on implementing recognizers *in* hardware: directly, on mobile platforms. Our goal in this chapter is to explain (1) the motivation for this direction of research and (2) the technical prospects for such custom silicon recognizers.

Let us begin by listing the two most salient assumptions that underlie most work on mobile recognizers:

1. Mobile platforms offer limited computational resources.

2. As a consequence of assumption (1), recognizers on mobile platforms must make one of two unavoidable decisions:

   - reduce the capability of a software-based recognizer hosted on this platform or

   - move some part of the recognition computation off this platform.

Assumption (1) is not controversial. The computational resources of the processor(s) in a modern smartphone, limited by a 3 W total power envelope, are significantly less than those of a 10 W netbook processor or a 45 W laptop processor or a 100 W enterprise blade-server processor. Of course, the capabilities of all of these processors, across all of these hardware form factors, continue to grow. The processors in recent smartphones are beginning to resemble simple versions of individual cores in high-end multicore systems (Yu and Rutenbar 2009). Nothing in the hardware business ever stands still.

Assumption (2) is more interesting – at least, to our hardware-centric sensibilities. It is certainly clear that a best-quality recognizer for a large vocabulary, continuous, speaker-independent recognition (LVCSR) engine fully occupies a single core on a modern enterprise-class blade server (e.g. of the type used in call center telephony applications). Thus, moving that 'best of class' recognizer directly, without modification, onto a typical smartphone processor will result in unacceptable performance. In the more likely worst-case scenario, this software version will be too complex to host on the mobile hardware. This has spawned two very different sorts of solutions, as noted above.

One direction for solutions is to explore how to fit a recognizer onto the reduced resources of the specific platform. This has given rise to a significant new research area, that of *embedded* (i.e. reduced-resource) recognizers. For example, we might reduce the vocabulary size or the complexity of the acoustic model, to better align what the recognizer needs with what the platform provides. We might also limit the scope of the recognizer to improve its usability; for example, voice selection of elements from menu lists can make powerful use of the limited dialog context to improve accuracy. The PocketSphinx work from Carnegie Mellon University (Huggins-Daines *et al.* 2006) and the Pocket-SUMMIT work from MIT (Hetherington 2007) are good examples of this work.

The alternative direction argues that best-quality recognition will never fit on mobile platforms, and must be hosted elsewhere. For handsets, which have built-in cellular or wireless communication, the solution is obvious: move the voice data to some place where it can be run on a best-quality recognizer. This idea gives rise to the work on *distributed* and *networked* recognizers, which move at least some part of the computation off the resource-limited local platform. At its simplest, we might handle the desire for dictation (e.g. to dictate an email) by sending the raw voice data to a high-capacity server at some remote location. Or we might host the recognizer in a more heterogeneous fashion: reduce the raw voice stream to a much smaller set of digital features, and send this digital data stream to a high-capacity server to complete the remainder of the recognition task.

Other chapters in this book discuss progress on both of these solution strategies. However, this chapter suggests a third alternative: *render the recognizer itself in hardware*. The idea is not outlandish. We note, for example, that video playback is now handled in dedicated hardware on essentially all high-end smartphones. What started as a novelty, implemented in software, has evolved to a necessity, implemented much more efficiently, in hardware. The operative question for the work discussed in this chapter is whether speech recognition will make the transition from 'nice to have' to 'must have' on these platforms, thus possibly justifying the hardware solution. Unfortunately, answers to such questions involve economics, which is not the topic at hand. Instead, we focus in this chapter on the technical side of this idea: *why* we might want to move speech recognition into silicon, *how* we might accomplish this and *what* results we might expect for such a transition.

Of course, novel solutions bring novel problems along with new opportunities. As a preview of ideas to come, let us briefly enumerate the most important of these.

The advantages of custom hardware are as follows.

- *Fewer performance compromises.* The hardware strategy significantly rebuts the assumptions we began this section with. We may no longer need to reduce the capabilities of the recognizer to fit the platform, nor send it off the platform (with the concomitant delay uncertainty and potential communication cost).

- *Best possible energy efficiency of any form factor?* As we shall see, the dedicated hardware solution is always the most energy efficient, measured in terms of computations performed per unit of power or energy expended.

- *Superior decoding speed for complex acoustic or language models.* Just as was the case with video playback, the hardware solution can be designed to offer superior recognition speed, even for relatively complex recognition tasks, while running on very little power.

The disadvantages of custom hardware are as follows.

- *Inverse relationship between flexibility and efficiency.* The dedicated solution is extremely efficient, in terms of speed, in terms of hardware size, in terms of power, etc. But the more single-purpose the engine, the more we sacrifice flexibility. Unlike software, we may not be able to simply 'reprogram' a hardware solution to do something different. This creates interesting research problems that seek to balance the tension between dedicated (and thus efficient) hardware function, and configurable (and thus flexible) function.

- *The first chip is expensive.* Although we mentioned above that economics was not our focus, we must nevertheless note that designing a dedicated chip – a so-called *application specific integrated circuit*, or ASIC – is an expensive proposition. For example Santarini (2008) estimates that in 2008 integrated circuit technology, the nonrecurring engineering costs to design a dedicated ASIC are roughly $20M. There are a variety of less expensive solutions, notably configurable hardware such as *field programmable gate array* (FPGA) technology (Kuon and Rose 2006; Kuon *et al.* 2008), but there is always a cost-versus-performance trade-off to be made. The usual strategy in this domain is to target chips that can be inserted into high-volume products, so as to amortize this initial cost over many units. Mobile phones are such extremely high-volume products.

We focus in this chapter on hardware-based recognizers for mobile phone platforms. This is partly in response to the need to amortize design cost over very large volumes of deployed chips. But this is also because the typical handset hardware platform offers the most challenging power, cost and size constraints.

One final caveat is worth mentioning. Descriptions of hardware-focused performance limits have a finite shelf life. Performance goals that seem lofty today may seem quaint five years from now. This is simply the nature of the business. (We note that the specific performance numbers used in our examples are from circa-2008 technologies.) Advances in semiconductor technology will continue to improve the processors in high-end smartphones. Multicore processors may soon make an appearance on these platforms (ARM Holdings 2009), offering new challenges in how best to parallelize speech tasks. But these same advances will also improve the performance and efficiency of dedicated hardware. Inevitably, the best partition between hardware and software will depend on the interplay amongst the application, the platform and the required performance and cost. One of our goals in this chapter is to try to explain the nature of these hardware-versus-software trade-offs, as they apply in the domain of speech recognition.

The remainder of this chapter is organized as follows. Section 2.2 begins with a brief overview of the capabilities and limitations of the various computational and memory components in a modern handset. Section 2.3 then profiles a modern software LVCSR recognizer to illustrate exactly how a real speech engines uses its hardware resources; this offers us the essential guidelines for what we need to optimize, to move this application directly into hardware. Section 2.4 then offers a brief review of conventional software approaches to recognition on resource-limited mobile platforms: reduced-resource embedded recognizers, network recognizers and distributed recognizers. Section 2.5 surveys some of the work we have done at Carnegie Mellon on the design of custom silicon recognizers. Finally, Section 2.5 offers concluding remarks and outlines some future directions for work in this area.

2.2 Mobile hardware: Capabilities and limitations

In this section we take a quick look 'inside' a representative mobile platform (circa 2008) to get a preliminary sense of what real hardware actually looks like. We use this as a launching point for a review of the processing, memory and power limitations that define a modern mobile handset, and a brief review of what Moore's law really means for the mobile hardware space.

2.2.1 Looking inside a mobile device: Smartphone example

With over three billion service subscriptions worldwide as of 2008 (Newman *et al.* 2007), the cell phone is the most ubiquitous example of a mobile hardware platform. From their brick-like beginnings in the 1980s, cell phones have evolved into the ultra-compact, ultra-lightweight forms we are accustomed to today, along the way subsuming many of the functions of other devices. Current cell phones may not only function as phones, but also as personal digital assistants (PDAs), email clients, web browsers, cameras, video recorders, and even televisions. Indeed, no other device has driven the development of mobile technologies as has the cell phone. The modern, feature-rich cell phone (or *smartphone*) is therefore a natural example for us to consider.

The specific example we shall take is that of the Samsung SGH-P920, a quad-band GSM cell phone featuring a 1.3 megapixel camera, video recording,

MP3 playback, *Bluetooth*™, stereo audio, and the capability to receive digital television (Carey 2007). This particular smartphone is of particular interest to us since it includes 'extra' functionality, in this case digital television reception, implemented with extra custom hardware.

Figure 2.1a and Figure 2.1b show the forward and reverse sides of the main SGH-P920 circuit board, with the major integrated circuits labeled. The remainder of the real estate on this board is dedicated to discrete components in support of the electrical needs of the labeled chips – for example, power supply regulators, filter capacitors and so forth. To give some idea of scale,

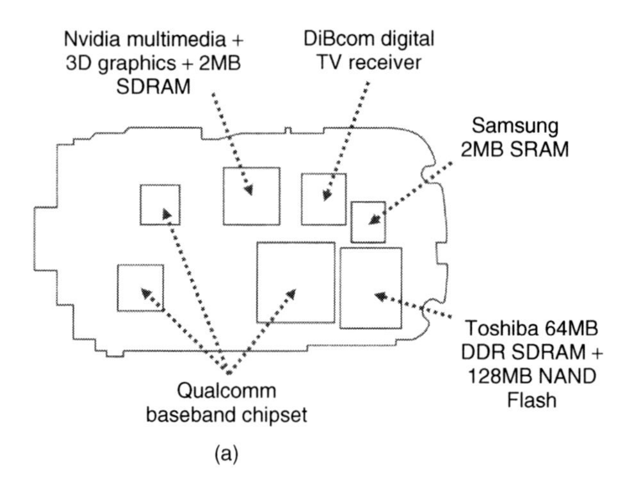

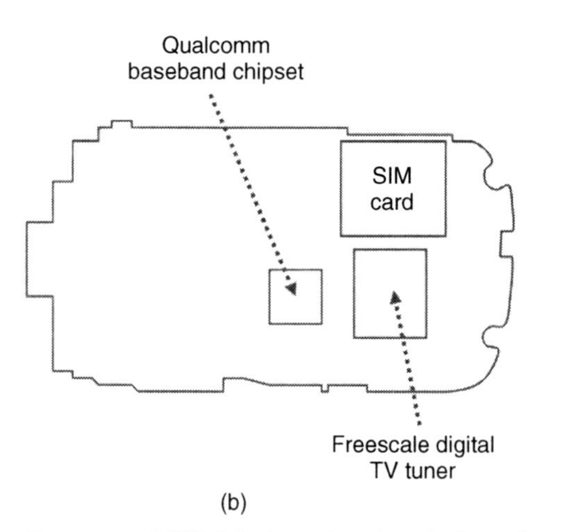

Figure 2.1 Samsung SGH-P920 main circuit board, redrawn from Carey (2007).

the footprint occupied by the SIM card slot is also marked. We now briefly describe the hardware resources offered by each of the labeled components.

Qualcomm baseband chipset. Most cell phones are built around a standard chipset provided by one of a limited number of vendors, and the SGH-P920 is no exception. In this case, basic cell phone functions are implemented using the Qualcomm MSM6250A baseband chipset, which consists of four individual chips. Features of this chipset are an ARM926EJ-S processor, which serves as the phone's primary processor, and two QDSP4000 digital signal processors (DSPs). The chipset provides the following functionality:

- both quad-band GSM and WCDMA telephony;
- MPEG-4 video encoding/decoding;
- JPEG encoding/decoding;
- 2D/3D graphics acceleration;
- a high-polyphony MIDI synthesizer;
- a 2 megapixel digital camera interface;
- support for GPS;
- *Bluetooth*™ connectivity;
- USB connectivity.

Nvidia multimedia processor. In order to provide higher performance audio and video, the Qualcomm chipset is supplemented by the Nvidia GoForce 5500 multimedia processor. This chip provides improved 3D graphics performance (sufficient for phone-based gaming), surround sound and support for displaying digital television on the phone. The same chip package also houses 2 MB of SDRAM on a separate die, which serves as working memory for the multimedia processor.

Freescale digital TV tuner. This chip is the first of two required to support digital television, and isolates incoming digital TV signals, which are then provided to the digital TV receiver. This chip is specifically designed for mobile applications, incorporating power-down and deep-sleep modes.

DiBcom digital TV receiver. This chip demodulates and decodes the digital TV signal provided by the tuner chip above, providing decapsulated audio and video streams to the Nvidia multimedia processor for display. Again, this chip is designed for low-power operation, consuming around 20 mW while processing a digital TV signal.

Samsung 2 MB SRAM. A separate 2 MB of SRAM serves as working memory for the DiBcom digital TV receiver chip.

Toshiba 64 MB SDRAM and 128 MB flash memory. Sixty-four megabytes of DDR SDRAM serves as the main working memory for the phone, and is shared between the baseband processor chipset and the Nvidia multimedia processor. Another multiple die package, this chip also contains 128 MB of NAND flash memory, which is used to store music, photos and video, among other data.

As we can see from this example, the resources available on a typical mobile device, while not insignificant, are not quite what one would expect to find on more conventional computing platforms. Rather than a single, general-purpose processor and a single, unified memory hierarchy, in mobile systems we see the use of multiple specialized processors and a range of memory types, tailored to the needs of the system. The reason for this is simple: computing tasks may be performed vastly more efficiently in hardware than in software. Indeed, due to size and power constraints, often the only way computationally expensive features (such as 3D graphics and digital television) can be implemented on mobile devices is through the addition of dedicated hardware (in this case, a TV tuner chip, a multimedia processor chip, and an additional separate SRAM memory chip). We devote the remainder of this section to detailing the constraints on this hardware, considering specifically processing resources, memory systems and overall power consumption.

2.2.2 Processing limitations

Processors found on mobile devices are designed to maximize performance for a given power budget or size, making them considerably less powerful than their workstation counterparts, which have significantly less onerous power and area constraints. For instance, ARM11 family processors (Cormie 2002) commonly found on mobile devices (circa 2008) usually run in the 200–600 MHz range and are capable of processing up to 740 million instructions per second (MIPS), much less than workstation processors, which run in the low-gigahertz range and can achieve over 22 000 MIPS. Furthermore, a larger variety of processors are found in embedded devices, leading to significant architectural differences that impact how speech recognition code should be written in order to optimize performance while still maintaining accuracy.

Two related processor parameters that particularly impact performance are the number of instructions fetched/decoded per cycle, and the number of functional units. If the processor fetches too few instructions per cycle there will be many functional units left idling while awaiting commands. On the other hand, if there are too few functional units, the processor will not be able to fetch more instructions as it is waiting for the current ones to finish. The number of instructions processed per cycle will vary significantly depending on the type of processor. Simpler mobile processors can only fetch and decode one instruction per cycle, while more powerful mobile processors have a superscalar architecture that permits the issue of multiple instructions per cycle. Some processors achieve this by extracting multiple operations from a single fetch (generally 32-bit), either by utilizing some 16-bit instructions in the case of ARM processors, or by issuing multiple operations from the same instruction for vector processors.

Different types of functional unit also have an effect on the number of instructions processed per cycle, and the execution time. Digital signal processors commonly have multiply-accumulate functional units and instructions, allowing what other processors complete in two sequential instructions to be completed in one. Many embedded processors do not have floating-point functional units, necessitating the conversion of all math from floating- to fixed-point. (Since fixed-point values are merely integers with an implied radix point, all floating-point operations may be substituted with integer operations, followed by a shift operation to align the radix point.) Although the use of fixed-point math does increase the number of instructions, fixed-point arithmetic is attractive even on processors with floating-point support, since floating-point instructions take more cycles to process than integer instructions. Also, more integer functional units may be available, so throughput is higher.

Without access to fast memory, a processor will not have the operands available to execute instructions. Unfortunately, in the mobile domain, the amount of on-chip memory is limited. Some processors will only have small level-one instruction and data caches and no level-two cache, making memory management an important consideration in order not to incur the long access latencies of off-chip memory. Some processors include a user-managed scratch-pad memory which can be populated with direct memory accesses from off-chip memory while the processor is running different instructions. This is especially useful in prefetching data for predictable memory accesses, so the data needed is always on-chip.

Embedded processors have also small translation lookaside buffers (TLBs) compared to workstation processors, making the size of the memory footprint a concern. When the processor attempts to access a memory location not found on a page in the TLB, an operating system interrupt is triggered that takes many cycles to process.

Conditional branches are particularly expensive, since most embedded processors have no branch prediction units and cannot execute instructions out of order, so must stall until the branch is resolved. While some designs have added special instructions to reduce the overhead of loops, in general it is more advantageous to increase the number of instructions and computations rather than add conditional branches.

Finally, while it is the case that no single processor specification determines how fast speech recognition might run, the processor clock frequency multiplied by the number of MIPS gives a good first-order estimate. Indeed, if all processors compiled the same code into the same number of instructions this heuristic would be exact. However, the number of instructions depends on the compiler and underlying architecture. For example, a DSP can complete multiply-accumulate operations in a single instruction, where other processors require two. We must therefore be careful in directly comparing MIPS in the mobile domain, since this does not necessarily equate to the amount of processing work undertaken.

2.2.3 Memory limitations

The memory system is another area in which there are significant differences between mobile and traditional computing platforms. In a traditional computer, we expect to see a unified memory hierarchy, where data is loaded from disk into main memory, into perhaps multiple levels of cache, finally to be consumed by the processor. The means by which this happens is largely transparent to the programmer – save for accessing a file on disk, data appears to reside in one large memory; it may just take longer to reach depending on location. Mobile devices differ in several ways.

Firstly, the components of a traditional memory hierarchy are often not even present in mobile devices due to size or power limitations. Hard disks tend to be large, and contain moving mechanical parts that consume relatively large amounts of power. With the notable exception of some media players, permanent storage will often take the form of non-volatile memory, such as flash.

Secondly, the different tasks performed by a mobile device will often have a wide range of different memory requirements. For instance, functions such as streaming video require high-bandwidth sequential access to memory. Flash memory may be required to store and play music, and will be frequently read, but not often written. A processor running Java applications will access memory more randomly, and therefore can take advantage of a more traditional memory system, perhaps with some sort of cache. Rather than attempting to build a single memory system that satisfies all these needs, it is far more efficient to design separate memory systems, each tailored to a particular task. We see this in the example of Section 2.2.1, which makes use of no fewer than *four* separate memory systems. Indeed, many mobile applications would not even be possible without this approach.

In order to meet the broad range of memory requirements seen on mobile devices, a number of different memory technologies are employed. We will now discuss each of these, describing the technology, its strengths and weaknesses and the scenarios in which it is applied in mobile devices.

SRAM

Static random access memory (SRAM), is the most straightforward type of memory to implement in a mobile system. It is referred to as static since, unlike some other memory types, it does not need to be refreshed to maintain its contents, providing power is maintained. Data and an address are supplied to the memory, the data is written, and it remains available to be read as often as necessary until that memory location is overwritten with new data.

SRAM stores each bit using (typically) an arrangement of six transistors. Four of these six transistors are used to form two cross-coupled inverters, which store a data bit. The remaining two access transistors are used to enable reading or writing of the stored bit. Collectively, this structure is known as a *cell*, and many cells are arrayed to form an SRAM of a given size and bitwidth. While arrangements other than the standard six transistor cell are possible, this is the most common.

SRAM is the fastest memory technology to read or write, with access times under 10 ns possible for high-speed applications. Such access times, however, come at the cost of power consumption, which may be as high as several hundred milliwatts while actually performing a read or write operation. Low-power SRAMs, specifically designed for mobile applications, may consume as little as 10 mW, but have access times in the 50–70 ns range.

SRAM may be present in mobile devices as either a separate integrated circuit, an additional die in the same package as another chip, or fabricated directly onto a chip itself. In the first two cases, SRAM sizes may range from 64 Kbit through to 64 Mbit. In the case of SRAM fabricated directly on-chip, size is determined by application, and the amount of die area one is willing to devote to memory. It is not unusual for custom integrated circuits to contain tens of small, on-chip SRAMs, perhaps of only a few thousand bits each. An excellent example of why one might want such an arrangement, for example, would be to store lookup tables containing data for a particular calculation performed by a chip. Larger SRAMs are fabricated directly on-chip for use as caches. The larger an SRAM is, the more power it will consume.

DRAM

The basic cell used to store a bit in a dynamic random access memory (DRAM) consists of only a single transistor and capacitor. The value of this bit is represented by the charge on stored on the capacitor, and as any capacitor left over time will lose its charge, periodically a DRAM must be *refreshed*. That is, its contents must be read and written back to restore the charge on the capacitor. This is why the memory is referred to as dynamic.

The chief advantage of DRAM is its high density, as each cell is very simple. A single DRAM chip may store up to 2 Gbit, an order of magnitude more than SRAM. Disadvantages of DRAM compared to SRAM include greater power consumption, slower access times, and increased implementation complexity.

Unlike SRAM, and due to the manner in which DRAM cells are read and written, the time to access a given DRAM location is not fixed. Rather, the time taken for an access will depend on the locations of previous accesses. Sequential accesses are fastest – this is known as *burst mode* – and DRAMs are well suited to providing data in this manner. Access times for a random DRAM location are in the range of 40–80 ns; after that initial access, sequential data can typically be provided every clock cycle for some limited number of cycles. Managing this complexity and the refresh process generally requires additional hardware in the form of a DRAM controller.

During a burst read or write operation, DRAM power consumption ranges from approximately 100 mW for low-power, mobile-specific parts, through to several hundred milliwatts for high-speed parts. Any DRAM controller would require additional power.

In spite of its increased complexity and power consumption, its large capacity and the ability to access sequential data rapidly make DRAM an excellent choice for many media applications, even in a mobile environment. In terms of speech recognition, for instance, DRAM is well suited to storing Gaussian mixture model data and the hidden Markov model (HMM) state probabilities referenced by modern recognizers.

Flash

Flash is the third type of memory commonly encountered in mobile devices. It differs from SRAM and DRAM most significantly in that it is non-volatile, meaning that data is retained even without power. This is achieved through the use of a special transistor equipped with an extra floating gate on which charge is stored (Masuoka *et al.* 1984).

There are two types of flash memory, called *NOR flash* and *NAND flash*, so named for how the floating gate transistor is connected in each. In NOR flash, the transistors are connected in parallel as in a NOR gate array, while in NAND flash they are connected in series as in a NAND gate array. The two types of flash also differ in how they are accessed. To a large degree, NOR flash behaves like SRAM, but with longer read access times, and much longer write access times (although burst-mode accesses are also often possible). NAND flash, however, is structured somewhat differently. It is divided into pages, typically ranging in size from 512 bytes to 4 KB, and may only be written to a page at a time. NAND flash reads are burst-mode, and have a long initial access time. These, and other differences between the two flash types are summarized in Table 2.1.

Table 2.1 Comparing NOR flash and NAND flash.

	NOR flash	NAND flash
Read time	100 μs	20 μs
Write time	100 μs	200 μs per page
Erase time	0.1–1 s	Milliseconds
Size	Up to 64 MB	Up to 2 GB
Lifetime	100 000 cycles	1 000 000 cycles

As can be seen from Table 2.1, NOR flash is best suited to applications in which it is predominantly read, whereas NAND flash is a good choice when large amounts of data must be written, particularly if random access is not required. In mobile devices, a common application of NOR flash would be to store the firmware for the device, since this data is accessed every time the device is used but is updated only infrequently, when the manufacturer releases a firmware update. If the NOR memory used is sufficiently fast, code may even be executed directly from it, a technique known as *execute-in-place*. NAND flash is used for storage of data such as music and photos. It is ideally suited to this purpose, since this data is more frequently written, accessed sequentially, and can take advantage of the higher capacity of NAND flash.

With regard to speech recognition, the large language models referenced by modern recognizers are an excellent candidate for storage in NOR flash, due to their large size and predominantly read-only nature.

2.2.4 Power limitations

Other than the obvious size and weight constraints, limits on power consumption impact mobile device functionality most severely. While improvements in processing performance, memory capacity and display technology have come rapidly in recent years, improvements in battery technology have not.

The energy density (measured in Watt hours per liter) of the most important rechargeable battery technologies is shown in Figure 2.2 (Tarascon and Armand 2001). Also given are the years in which each technology first appeared commercially. We see from this figure that during the last forty years, while processing performance has doubled every two years, battery capacity has improved by only a factor of three.

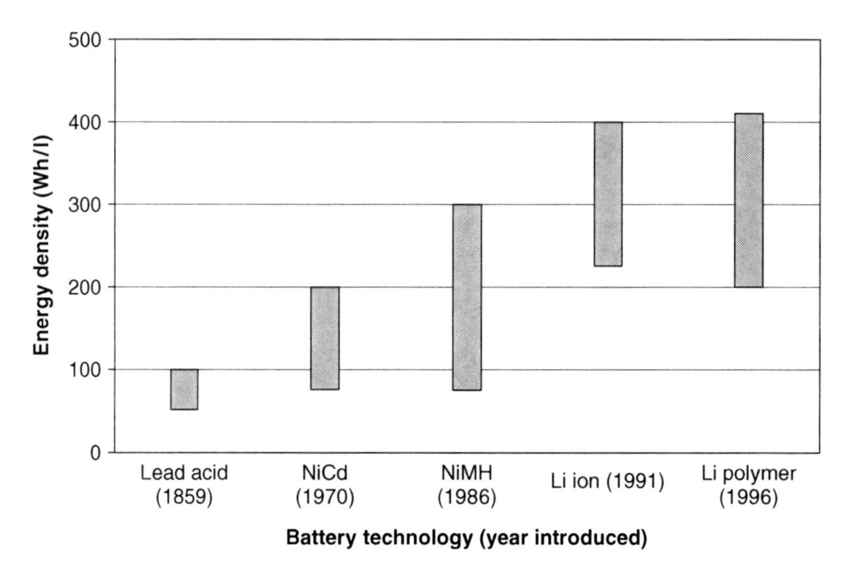

Figure 2.2 Energy density for common rechargeable battery types.

Of course, ever more powerful mobile devices continue to appear, each generation sporting a host of new features compared to the last. Since we cannot call upon improvements in battery capacity to power these, the burden has fallen on designers to greatly reduce power consumption, and indeed power management in modern mobile devices has become quite sophisticated. Any new feature appearing on a mobile device, such as automatic speech recognition, must be implemented given rather strict limits on how power consumption may be affected. In the following, we try to give some idea of what those limits are, again taking the cell phone as our example.

Typical cell phone designs have power budgets of around 3 W (i.e. this is the total power consumption of all the phone's components taken separately). Battery capacities range between about 700 mAh and 1300 mAh, at 3.7 V for Li-ion cells. Many of the subsystems within the phone, however, operate on voltages less than that of the battery, low-power integrated circuits typically requiring a 1.2-V or 1.8-V supply. Since converting to these lower voltages can be quite inefficient (depending on the method used), a non-trivial fraction of this initial power budget is lost immediately (Ambatipudi 2003).

Power consumption for the most common cell phone components is roughly as follows. Receiving and transmitting represent by far the greatest draw on power. During conversation, receiving and processing the incoming signal requires about 250 mW. Transmitting typically requires about 300 mW, but this could increase to 1.3 W or more if the distance to a cell tower is particularly large. The majority of the transmit power is consumed by the power amplifier responsible for driving the antenna (Szepesi and Shum 2002).

Of the remaining phone components, a color display draws between 250 mW and 350 mW during operation, depending on size (Toshiba 2009). Most of this is required for the display LED backlight, with the LCD panel itself only drawing about 15 mW. The speaker and its associated amplifier consume tens of milliwatts, depending on volume and the nature of the audio. Of particular interest for speech recognition applications, microphone power consumption is only 1–2 mW during use (National Semiconductor 2009).

The remainder of a cell phone's power budget (several hundreds of milliwatts) must be sufficient for all the other features one might want, including camera support, *Bluetooth*™, high-performance graphics and video processing, television and speech recognition. Therefore, in terms of power consumption, any additional feature requiring more than about 100 mW is difficult to justify; the impact on talk and standby time is generally unacceptable.

This leads us to the ways in which we can conserve power in mobile devices. One of the most effective of these is also the most obvious: turn off parts of the device that are not in use. While this behavior is readily apparent in components such as the display backlight, it extends much further. During pauses in conversation, for instance, speaker and microphone amplifiers will enter into standby modes, waking up again as soon as audio is detected.

Since transmitting requires the largest chunk of a cell phone's power budget, this is another area in which we might look to conserve power. Traditionally, the power amplifiers that feed a cell phone's antenna are optimized to achieve peak efficiency when operating at maximum power, about 28 dBm. Most cell phone conversations, however, take place sufficiently close to a tower that only about 10 dBm is required. Unfortunately, at these reduced power levels, power amplifier efficiency may drop from 40% to as little as 10%. Newer designs, based on InGaP (indium gallium phosphide) rather than GaAs (gallium arsenide) semiconductor technology, are able to achieve efficiency levels from 20% to 45% at peak (Miller 2006). Particularly at lower power levels, this is significant.

Improving the efficiency of voltage conversion between the battery and the various cell phone components has also been of recent interest. In the past, this has been achieved using low-dropout regulators, or LDOs. The efficiency of LDOs, however, falls rapidly as the voltage spanned by the LDO increases. As the voltages required by many low-power components (particularly integrated circuits) have dropped, there are therefore greater power savings to be had. Switching regulators designed specifically for mobile devices have therefore become popular, and can achieve efficiencies of up to 90% (Ambatipudi 2003; Szepesi and Shum 2002).

There are fundamental limits, however, to how far we can reduce the power required for signal transmission, or improve the efficiency of power conversion in mobile devices. We therefore turn our attention to the integrated circuit technologies used to implement the bulk of any device's functionality.

Power consumption in integrated circuits may broadly be divided into two components – *switching power* and *leakage power*. Switching power is that consumed when a transistor changes state, and until recently has accounted for the majority of power consumed by integrated circuits. Leakage power is constant, and results from unintended current flow between the power supply to a transistor and ground, modern transistors being far from ideal devices. As successive generations of integrated circuit technology have managed to cram ever more, increasingly non-ideal transistors on a chip, leakage power has begun to rival switching power in magnitude.

In mobile systems, power consumption in integrated circuits is controlled through a number of means (Chandrakasan and Brodersen 1992). Firstly, we try to run at low clock speeds, since this means less transistor switching (but also less performance). Secondly, we can apply the same obvious idea employed at the system level, and disconnect parts of the chip from the power supply when not in use. Finally, techniques such as dynamic voltage and frequency scaling can be used to adjust the performance to the bare minimum needed for an application, again saving power (Burd *et al.* 2000). Such techniques are well-suited to automatic speech recognition, where we generally require only sufficient performance to meet a real-time constraint.

2.2.5 Silicon technology and mobile hardware

For roughly forty years, Moore's law (Moore 1965) was in force for semiconductor technology. Simply put, we could expect on a regular basis (approximately every two years) to see transistor density and performance doubled. As a practical consequence, it was the always the case that the next computer chip was more capable, faster and available with more memory.

A few years into the 21st century, however, new microprocessors began to fall off this historical trend, as shown in Figure 2.3. Given the realities of nanoscale semiconductor devices at this time, it became possible to pack

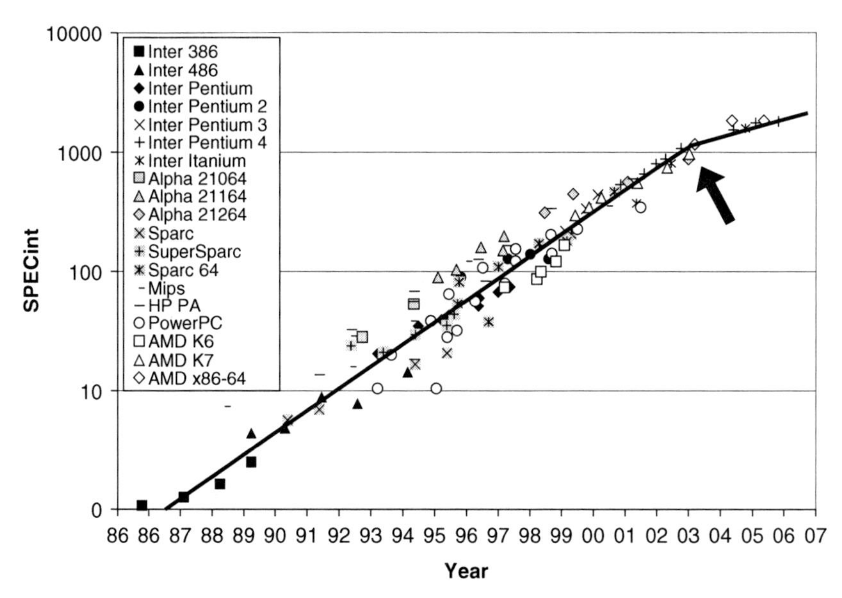

Figure 2.3 Performance scaling has slowed for processors in recent years (Horowitz, pers. comm.; see also Horowitz and Dally (2004). The vertical axis measures uniprocessor performance in terms of SPECint (Henning 2006). Individual processors appear as bullets. Around roughly 2003–2004, uniprocessor performance saw a significant inflection point.

twice as many transistors in a given area using the next technology node, but not with similarly improved speed. In other words, it became impossible to continue to deliver improved performance simply by delivering improved (greater) megahertz or gigahertz. It was at this point that we entered the so-called *multicore era*, and began to see individual microprocessors with separate parallel computing cores, each running at a few gigahertz. When it became impossible to deliver more performance by delivering a *faster* processor in the next technology generation, we addressed this problem by delivering *multiple* processors on the same chip in the next technology generation.

Although the most visible microprocessors in this Moore's law scaling story are the desktop, laptop and enterprise chips, there is still significant impact in the mobile space. Performance (MHz) has continued to improve for the most capable smartphones, for example, but has yet to reach the 3–4 GHz speeds of larger computers. It is not clear if the more onerous power limitations of the platform will allow individual processors to reach much beyond 1 GHz. As of this writing, multicore designs are suddenly under consideration for soon-to-appear generations of smartphones (Texas Instruments 2009). For example, the ARM Cortex-A9 MPCore architecture supports up to four individual processor cores, with floating-point arithmetic and with separate data and instruction caches per core.

However, it seems inevitable we shall soon face the same problem as the larger desktop and enterprise applications: per-core speed will stop increasing, and we shall need to deliver improved performance with more processing cores. This is particularly challenging for our speech-recognition applications. It is already the case that these are significantly less capable than their enterprise-class cousins. If per-core performance stops improving in the near future, what shall we do? We could, of course, work to parallelize the low-level functions of a single recognizer across a few cores. This is an exciting research topic, but a relatively new one, with some promising but immature results, for example Cardinal *et al.* (2009) and Parihar *et al.* (2009).

Another avenue of attack is to stop relying solely on software to deliver the recognizer. This is already the path taken for video playback on modern mobile handsets. A central observation of this chapter is that software-only solutions are inevitably limited by the severe constraints on the capabilities of the hardware in any modern mobile platform. To be sure, those capabilities are improving, but equally certain is that Moore's law is slowing down, and forcing significant changes to new architectures. Moving functionality away from software and directly into hardware is the strategy we advocate in this chapter.

2.3 Profiling existing software systems

To understand how the hardware in mobile systems affects the applications that run on it, one must first understand exactly where the application spends its time and what hardware resources have the greatest effect on performance. This is especially important in the mobile space because each mobile device may have a different processor, memory storage or battery. In this section we profile a typical LVCSR speech recognizer on a variety of embedded processor configurations and measure execution time, resource usage and power consumption (Yu and Rutenbar 2009). From these results we determine which resources have the largest effect on performance, and some general guidelines on how to optimize performance.

2.3.1 Speech recognition overview

As depicted in Figure 2.4, speech recognition can be generally divided into three stages: *feature extraction*, *feature scoring* and *search*.

During *feature extraction* the input analog speech signal has its acoustic information distilled into a *feature vector*. First the analog signal is sampled

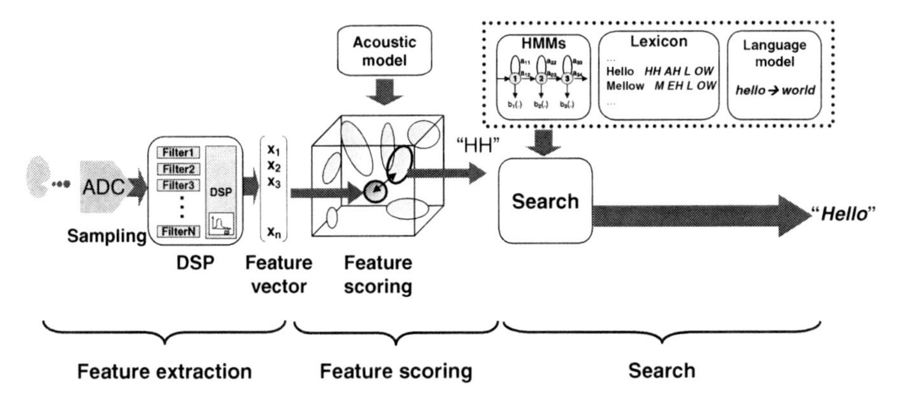

Figure 2.4 Speech recognition flow.

and digitized, then segmented into time intervals called *frames*. Various signal processing algorithms are then applied to produce features that represent the acoustic information of each frame. For each frame the acoustic features of its neighboring frames are combined to produce a feature vector. *Feature scoring* finds the probability of pronunciation for each unique atomic unit of sound by comparing the acoustic information parameterized in the feature vector with a model of each unique atomic unit of sound. The first two stages of speech recognition find the probability of each sound per frame, and the final *search* stage finds the most probable sequence of words formed from the sound probabilities. The final step includes a lexicon including the spelling of the recognizable words and a language model to indicate probabilities of certain word sequences.

There are a variety of implementations for each stage. Common acoustic features extracted from feature extraction include mel-frequency cepstral coefficients (MFCCs) (Davis and Mermelstein 1980) and linear predictive coding (LPC) coefficients (Hermansky 1990). To calculate feature vectors many use the acoustic features and their time derivatives, or a linear combination determined via principal components analysis (Jankowski Jr. 1992) or linear discriminant analysis (Brown 1987). Atomic units of sound can be represented by codebooks of Gaussian or Laplacian densities (Rabiner *et al.* 1985). In order to find the most likely sequence of sounds, one could use simple methods like dynamic time warp (Itakura 1975) or more computationally expensive HMMs with Viterbi beam search (Baker 1975).

2.3.2 Profiling techniques summary

We chose to profile Sphinx 3.0 (Placeway *et al.* 1997), an open-source large vocabulary continuous speech recognizer from Carnegie Mellon University.

In its most common configuration it computes 13 MFCCs for every 10 ms of speech and generates a 39-dimensional feature vector using the MFCC and its first and second time derivatives. Tied-triphone states are represented with Gaussian mixture models, and the search stage is a single-pass flat lexical search. While there are many other speech recognizers with alternative algorithms, Sphinx 3.0 contains the major elements of speech recognition and provides a good guide to the parameters to which speech recognition performance is most sensitive.

The speech corpus used in these experiments was the Wall Street Journal (WSJ) 5000-word task (Paul and Baker 1992). The acoustic model possessed 4147 tied-triphone states, each of which was represented by a Gaussian mixture model of eight mixture components, for a total of 33 176 Gaussians. Three-state HMMs were used to represent triphones, and the language model consisted of 4989 unigrams, 1.64 million bigrams, and 2.68 million trigrams. The test set used contained 40 min of speech and Sphinx 3.0's word-error rate was 6.707%.

In the mobile domain, important performance considerations include processing time, memory accesses and power consumption. Thus we profiled Sphinx 3.0 using these three metrics to determine how the processor architecture affects performance.

To measure processing time and memory accesses, we used SimpleScalar (Burger and Austin 1997), a cycle-accurate simulator for program performance analysis. SimpleScalar allows the user to specify processor attributes such as cache sizes, memory latencies and number of functional units, and then simulate programs on the processor architecture to collect performance statistics. This allows one to measure the impact of the different components of a chip to determine bottlenecks. We chose to model the ARM11 processor running at 500 MHz (ARM Holdings 2007; Cormie 2002) with a floating-point coprocessor to simulate the computing resources on a powerful mobile device. (We note additionally that in order to be able to run Sphinx 3.0, we needed a full floating-point capability on this simulated hardware.) Important SimpleScalar values in the baseline processor configuration include: 16 KB four-way associative level-one instruction (IL1) and data (DL1) caches with FIFO replacement, no level-two cache, an in-order core, a 128-entry bimodal branch predictor, four-instruction fetch per cycle, two-instruction decode, single-instruction issue, two integer arithmetic logic units (ALUs), single-integer multiply, two floating-point ALUs, a single-floating-point multiplier, and memory-access latencies of one cycle for level-one caches and 24 cycles for main memory.

Total power consumption is the sum of the dynamic and static (leakage) power of the CPU. To estimate dynamic power within the CPU, we used 65 nm CMOS technology numbers based on BSIM3 transistor models (Zhang *et al.* 2003) and Wattch (Brooks *et al.* 2000), an architectural-level power analysis tool that runs on top of SimpleScalar. For every clock cycle, Wattch monitors the instructions processed by each component and the energy consumption required to process instructions. For our statistics we used aggressive conditional gating, whereby the unused components are disabled to consume less power. While Wattch will give a good estimate of dynamic power, measuring the static power of a processor is very difficult because leakage is extremely sensitive to temperature, which is constantly fluctuating and cannot be easily measured. To estimate the static power of the CPU we multiplied the dynamic power consumption from Wattch by the ratio of static to dynamic power at ambient inside box temperature (45°C). For 65-nm CMOS technology, this ratio is 30% (Chapparo *et al.* 2004).

2.3.3 Processing time breakdown

As seen in Figure 2.5, 2% of the processing time is spent in feature extraction, 66% in feature scoring, and 32% in search. This is consistent with results from other recognizers, since feature extraction comprises digital signal processing algorithms that require very few operations and memory accesses to process audio rate (i.e. slow) input data. Feature scoring takes a long time because it needs to read in a large acoustic model and perform many computations. Although search consists of very little arithmetic, it is constantly accessing the active HMMs and language model, so needs lots of memory bandwidth. Any performance optimization must improve the feature scoring or search, and so the rest of this section will focus on these two stages.

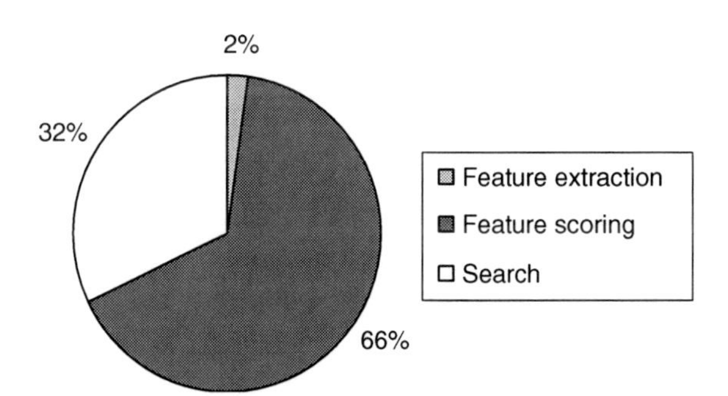

Figure 2.5 Time breakdown of Sphinx 3.0 on Wall Street Journal 5000-word corpus.

In order to better understand the bottlenecks, we also evaluated Sphinx 3.0 with a combination of ‘perfect’ memory, ‘perfect’ instruction fetch/decode/issue, and ‘perfect’ ALUs. By removing these constraints we can see which of the parameters performance is most sensitive to. To achieve 'perfect' memory we configured SimpleScalar to have one-cycle latencies for all cache and off-chip memory accesses. For 'perfect' instruction fetch/decode and ALUs, we just set the number of instructions fetched/decoded and number of ALUs to the largest values SimpleScalar would accept (sixty-four fetches, sixteen decodes, eight integer ALUs, eight integer multipliers, eight floating-point ALUs, eight floating-point multipliers). Figure 2.6 shows the cycle count of the three stages and the total time normalized against the baseline model results.

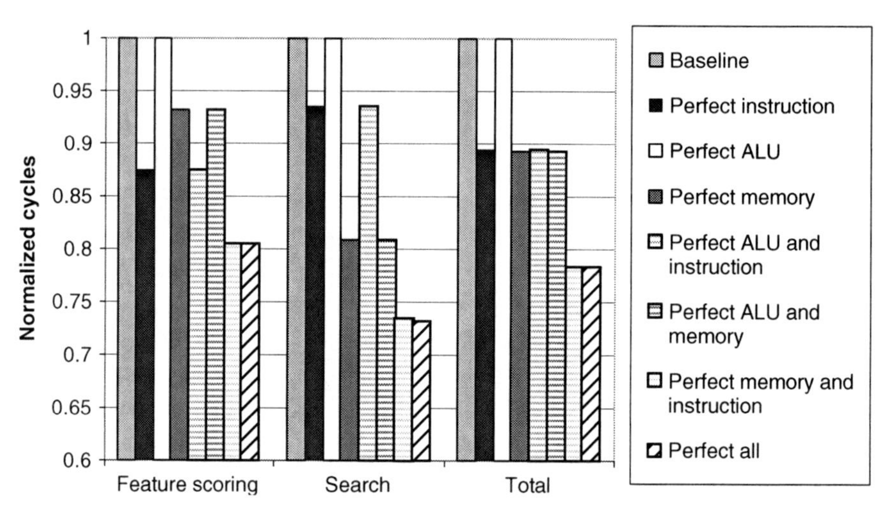

Figure 2.6 Effects of 'perfect' memory, ALU, and instruction on execution time. Note that y-axis starts at 0.6.

Memory and instructions have the greatest impact on execution time. It is intuitively obvious why memory is important, because both feature scoring and search require reading amounts of data that are too large to store on-chip. However, instructions generally do not limit performance, especially in a program with as much parallelism as speech recognition. In this case, the ARM11 was designed to be scalar and in-order, which limits the instruction-level parallelism that can be exploited. Without enough instructions being issued the functional units remain idle and so 'perfect' ALU shows no improvement. When examining individual stages, one finds feature scoring benefiting more from instructions fetched/decoded because of its inherent parallelism, while search favors faster memory as it has less operations but requires more memory bandwidth. For the WSJ 5000-word corpus, improving memory has a greater impact on overall performance, but for other test sets that spend a greater fraction of execution time in feature scoring, run time will benefit more from using a superscalar processor.

2.3.4 Memory usage

Speech recognition performance is very sensitive to memory bandwidth so it is important to know the memory usage breakdown. This includes both static and dynamic memory usage, which refer to the memory needed to store the models and memory dynamically allocated during runtime, as well as the memory bandwidth consumption. These results are not very sensitive to processor configuration because memory usage is a function of the algorithm and not the hardware.

The static and dynamic memory needed for feature scoring and search are shown in Table 2.2. Static memory requirements are not critical, as even mobile devices have large amounts of memory to store the data, and reading the data to initialize the models during run-time is not considered part of execution time. On the other hand, dynamic memory usage is very important because this represents data usually stored in off-chip memory that must be read to the processor and supplied to the ALUs. Also, increasing dynamic memory results in more TLB misses, which may require waiting hundreds of cycles for the operating system to resolve.

Table 2.2 Static memory and dynamic memory usage for WSJ 5000-word corpus.

	Feature scoring	Search	Total
Static	10.5 MB (23.2%)	34.7 MB (76.8%)	45.2 MB
Dynamic	13.6 MB (33.5%)	27.0 MB (66.5%)	40.6 MB

The DRAM access breakdown is shown in Figure 2.7. Although feature scoring takes more than half the execution time, it makes up only a small

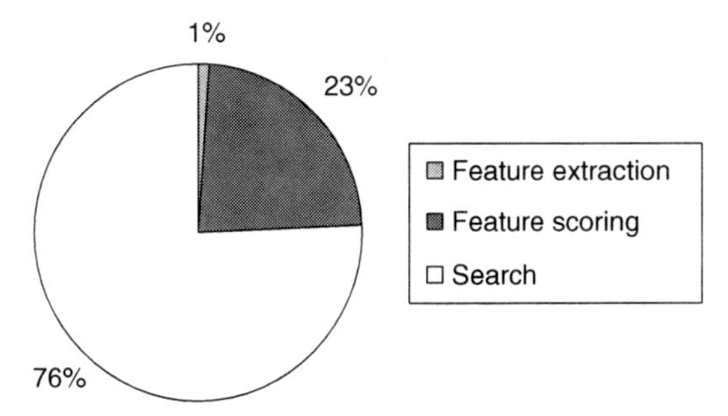

Figure 2.7 DRAM access breakdown of Sphinx 3.0 on Wall Street Journal 5000-word corpus.

percentage of DRAM accesses. When one considers the total amount of time it takes to read the data from DRAM, feature scoring is an even smaller percentage. The addresses of feature scoring accesses are sequential in memory, so the data can be continuously burst from DRAM, with corresponding burst-mode latencies. While some search accesses are sequential in memory, such as reading active HMMs, language model accesses are much more random and require multiple pointer dereferences, incurring slower random-access latencies. For real-time recognition of the WSJ 5000-word corpus the processor requires a DRAM bandwidth of 1.51 GB/s.

2.3.5 Power and energy breakdown

Since the methodology used to estimate leakage power is fairly crude, the exact power measured during our simulation study is of less interest than the relative breakdown of where the power is being consumed, which is more accurate. Hence in this subsection we focus on these relative breakdowns.

Equally important when measuring the impact on battery life, energy (measured in Joules) is a better metric than power (measured in Watts, where 1 Watt is 1 Joule/second). Hence, in Figure 2.8 we shown the breakdown of energy consumed for the three stages. Power is the energy dissipated divided by time, and can be used to compare different programs only when the execution time is the same. If one program consumes lots of power but runs very briefly (e.g. feature extraction) it will drain the battery more slowly than an application that consumes less power but runs for long periods of time (e.g. search). Using power to compare the two programs would lead one to optimize the power consumption of the first program but improving the second would have a bigger boost to energy savings.

Feature scoring consumes a greater fraction of the energy than its fraction of the execution time because the processor is constantly computing and has

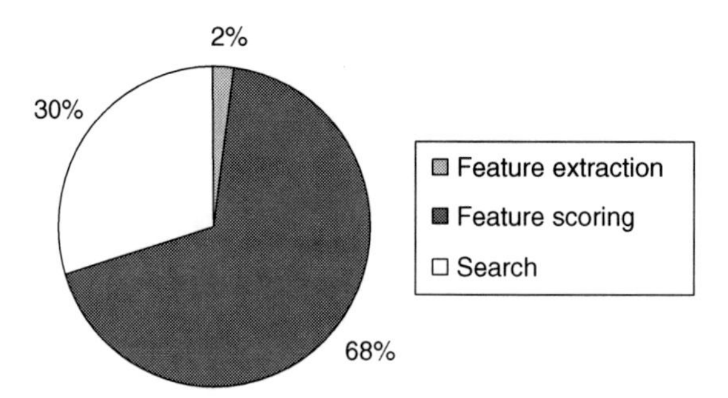

Figure 2.8 Energy dissipation breakdown of Sphinx 3.0 on Wall Street Journal 5000-word corpus.

little idle time. Conversely, during search there are many processor components that can be turned off while waiting on memory, so the energy consumed per cycle is lower.

Adding more functional units or increasing the size of the caches will generally increase the power consumed per cycle, but depending on the decrease in execution time there may be energy savings. In Figure 2.9, for different IL1 and DL1 cache sizes, we measured execution time, power dissipation and energy consumption relative to the baseline configuration. In both graphs we see that increasing the cache size increases the power consumed, which makes intuitive sense because there are more gates. In terms of energy, in Figure 2.9a the energy dissipated monotonically increases with cache size, but in Figure 2.9b there is local minimum for an 8-KB DL1. Increasing IL1 does not reduce the execution time enough to overcome the extra power dissipated per cycle, but increasing DL1 does. This shows that for optimal battery life one should not necessarily choose the smallest value for every parameter because energy dissipated does not always monotonically decrease.

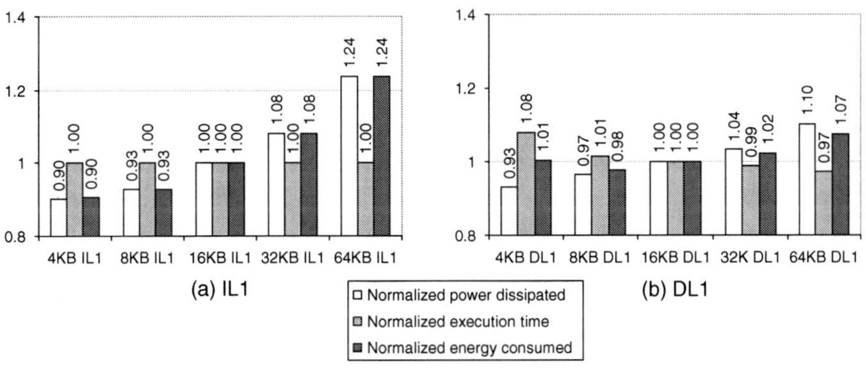

Figure 2.9 Normalized power, execution time, and energy for varying IL1 and DL1 cache sizes.

Further analysis helps explain the difference between the relationship of cache size to energy consumption for the two types of caches. A 4 KB IL1 already has a low miss rate of 0.19%, so increasing IL1 size hardly improves the miss rate or execution time. The algorithms that take the most time in speech recognition, the Mahalanobis distance computation and Viterbi beam search, simply do not require many instructions to implement. Thus IL1 size does not impact execution time and a larger IL1 just increases power dissipated and energy consumed. Conversely, total execution time dramatically decreases when increasing the DL1 size from 4 KB to 8 KB. The sharp descent is due to improvement in the feature scoring stage. Analysis of the assembly code shows the extra DL1 misses occur because the working set for computing tied-triphone state probabilities cannot exist in a 4 KB four-way cache without conflict. In larger caches the feature vector can be read in without displacing the Gaussian constants, but not so in a 4 KB DL1, leading to cache thrashing and poor performance. This underscores the importance of sizing the Gaussian mixture models – at least for a software implementation – such that tied-triphone state probabilities can be computed without excessive DL1 misses.

2.3.6 Summary

Clearly, a well-designed speech recognizer should be optimized for the available hardware resources. By profiling Sphinx 3.0 we now understand the characteristics of the three stages of speech recognition and have also determined the parameters to which performance is most sensitive. Although different speech recognizers, different acoustic/language models, and different hardware will change the specific breakdown percentages, the general trends and key characteristics to pay attention to still hold true.

When either designing a speech recognizer for a particular system or choosing a system for a specific recognizer, it is important to find the right balance of memory bandwidth, functional units, and instructions fetched/decoded/issued to optimize performance. For example, for our baseline system, increasing the number of functional units did not improve execution time and only increased energy consumed because there were not enough instructions fetched/decoded. In other systems the scenario could easily be reversed, with too many instructions ready to be issued but not enough functional units. The results also show that the size of the data cache relative to the working set can greatly impact both execution time and energy consumed. In most systems speech recognition will probably be most dependent on the available memory bandwidth, as this affects the two stages that make up most of the processing time (feature scoring and search). Increasing memory bandwidth will also improve energy performance, as there will be less leakage power during cycles when the data-starved processor is idling. Finally, in our in-order, scalar ARM11 simulations, we found that limited instructions issued per cycle also limited performance. More advanced processors have architectural differences such that instructions are less of a bottleneck.

2.4 Recognizers for mobile hardware: Conventional approaches

So far in this Chapter we have described typical mobile hardware platforms, and profiled the Sphinx 3.0 recognizer in order to gain some understanding of the hardware resources required of a 'best-quality' LVCSR speech recognition engine. In this section we discuss the implications of mobile hardware for three important classes of conventional mobile speech recognition systems – embedded, network and distributed. We use these to establish some useful context for our own non-conventional hardware-based solution.

2.4.1 Reduced-resource embedded recognizers

An embedded speech recognizer must complete the entire speech recognition process in roughly real-time on the limited resources provided by the mobile platform. The conventional solution is to reduce the computational complexity – and thus the required hardware resources – of an LVCSR recognizer, making it 'fit' on the smaller platform. Speech recognition performance is sensitive to clock frequency, memory bandwidth, instructions fetched/decoded and functional units, and in each case an enterprise-class workstation processor is much more powerful than the mobile processor. While the exact difference in performance depends on the specific mobile processor, two speech recognizers that achieved real-time recognition on embedded devices ran at around 0.05 times real time on a Pentium 4 system (Hetherington 2007; Huggins-Daines *et al.* 2006). These numbers give a rough sense of how far these recognizers had been 'reduced' to fit a mobile platform. Putting it another way, these reduced-resource versions, when run on the enterprise hardware, were roughly 20 times faster. To manage this hardware-resource gap, embedded speech recognizers have made algorithmic changes in order to meet real-time requirements with an acceptable decrease in recognition accuracy. Other chapters in this book discuss these sorts of embedded solution, but here we briefly highlight some embedded recognizers and their algorithms (see Table 2.3).

Fixed-point arithmetic and quantization

Most embedded speech recognizers use fixed-point arithmetic instead of floating-point arithmetic. Since floating-point notation has much larger range, this means the radix point must be carefully positioned. Placing too few bits before the radix point will cause excessive overflowing, while too few bits after the decimal place will decrease the resolution and precision making it difficult to discriminate between values.

Using fixed-point arithmetic also requires representing the acoustic-model parameters in fixed-point notation. In practice, most recognizers do not use the full register size of a machine (generally 16 or 32 bits) to store a single value, and instead use aggressive quantization schemes to represent values in as few bits as possible. For example, if the means were only allowed to take on 16

Table 2.3 Comparison of embedded speech recognizers.

	PocketSphinx Huggins-Daines *et al.* 2006	PocketSUMMIT Hetherington 2007	AT&T Bocchieri and Blewett 2006
Fixed-point	Yes	Yes	Yes
Corpus	Resource mgt	Jupiter weather	Resource mgt
Vocubulary size	1000 words	2000 words	1000 words
Processor	206 MHz StrongARM	408 MHz ARM	206 MHz StrongARM
Decoding speed	0.87 × Real time	Real time	Real time
Dynamic footprint	–	3.2 MB	7.5 MB
Language model	Bigrams	Trigrams	Word-pair grammar
Other	Single codebook	Finite-state transducers	Finite-state transducers

unique 32-bit values, then it would only take four bits to represent each mean in memory. Although decoding the actual value will take an extra lookup, the memory bandwidth required to read the means decreases by a factor of eight. Quantizing the acoustic model is especially beneficial because it decreases the amount of memory bandwidth and the dynamic memory footprint. Generally, the number of bits used to represent values is a power of two for byte-alignment purposes. While it is possible to compress the data to bitwidths that are not powers of two, this would require extra shifting and tracking where the mean ends (Bocchieri and Blewett 2006).

Some recognizers replace part of the Mahalanobis distance calculation with a single table lookup through aggressive quantization (Hetherington 2007; Vasilache *et al.* 2004). By restricting the difference of the feature vector and mean to 2^{α} values and the variance to 2^{β} values, there will only be $2^{\alpha+\beta}$ unique Mahalanobis distance values. These can then be precomputed and stored in a lookup table, indexed by the difference of the feature vector and variance concatenated together. Thus a table lookup replaces two multiplications, although steps to limit the range of the variances, such as variance normalization, may then be required.

Mixture component selection

When probabilistic mixture models are used to represent atomic acoustic units, the overall probability is usually dominated by a few of the mixture components. To reduce the number of needless computations, many mobile speech recognizers use mixture component selection techniques, such as context-independent Gausian mixture models (GMMs) or Gaussian shortlists (Franco *et al.* 2002). In some cases, speech recognizers will completely skip over a frame and reuse probabilities from the previous one if the feature vector is similar enough (Hetherington 2007). This reduces the memory bandwidth requirement, but also means the data accessed will not always be continuously stored in memory, limiting the advantage of the DRAM burst mode. This also requires extra instructions to handle the probability backoff for cases where atomic acoustic units have no mixture components scored.

Language model reduction

Reducing the size of the language model will also decrease the memory footprint and reduce bandwidth. The number of n-grams can be first smoothed using deleted interpolation then pruned using thresholds (Zhu *et al.* 2006). Like quantizing the acoustic model, language model weights can also be quantized, with the same benefits (Hetherington 2007). Finally, search can be divided into two passes, where the first pass uses a smaller language model in order to narrow the search space, and then a larger language model is used to re-score the lattice (Franco *et al.* 2002; Hetherington 2007).

2.4.2 Network recognizers

The key concept behind network speech recognition is to circumvent the resource limitations of mobile devices by moving the entire recognizer off the device, or *client*, and onto a remote *server*. Speech is transmitted to the server using the conventional speech coders employed in mobile telephony, where it is recognized and the result transmitted back to the mobile device. Two approaches may be taken by the server in processing coded speech. The first is to simply decode the speech before passing it into a standard speech recognition system. The second is to directly estimate features from the coded speech without first decoding it. This has advantages in terms of computational complexity and, perhaps less obviously, accuracy (Kim and Cox 2001; Pelaez-Moreno *et al.* 2001).

There are two main advantages to the network approach. The first of these, and the most important from a mobile hardware standpoint, is that nothing extra is required. The mobile device (presumably a cell phone) simply transmits speech exactly as it would for a normal voice conversation, using exactly the same hardware. Of course, if the device in question is not a phone, adding the necessary hardware just for speech recognition would probably not make sense.

A slightly less obvious advantage of the network approach, however, is that the speech recognizer is completely decoupled from the mobile device. The recognition system running on the server can therefore be upgraded or replaced without any disruption to the client, and may be extremely sophisticated.

There are, however, several effects that can impact the accuracy of network speech recognition. Firstly, speech is distorted during transmission between client and server. This is because the speech codecs used in mobile telephony run at low bit-rates and are optimized for human understanding, not computer recognition. For instance, such codecs tend to be based on LPC coefficients, which model the production of speech. Compare this with the MFCCs typically used for speech recognition, which model the perception of speech. While directly estimating features from coded speech can mitigate this mismatch, such an approach ties the recognition system used to a particular speech codec (Lilly and Paliwal 1996).

Furthermore, packet loss during speech transmission can also significantly degrade recognition accuracy (Mayorga *et al.* 2003). Since most speech recognition systems operate on a window several frames of speech wide, the loss of a single packet may introduce multiple errors. Also, the lower the bit-rate of the speech codec used, the more speech data will be lost per packet.

Finally, the major drawback of network speech recognition from a hardware perspective is simply the power required to transmit speech to the server, transmitting generally being the most costly operation a mobile device can perform in terms of power consumption.

2.4.3 Distributed recognizers

Distributed speech recognition systems are a cross between embedded and network systems, in that the recognizer front-end runs on the mobile device (client), but the computationally intensive back-end resides on a remote server. This solves the speech coding issues associated with network speech recognition, since only feature vectors are sent by the client.

Of course, it is still vital to the accuracy of the system that feature vectors are transmitted accurately between client and server. There are three stages in this process to consider: source coding, channel coding and error concealment (Tan *et al.* 2005).

The purpose of source coding is to compress feature vectors for transmission over limited bit-rate channels. Several techniques have been studied for doing so, and include scalar quantization, vector quantization (Digalakis *et al.* 1999) and transform coding (So and Paliwal 2006). Generally, source coding is lossy. It is also important to consider the effects of source coding on robustness, since different techniques will yield signals with different sensitivities to transmission errors.

The second stage in transmitting feature vectors between client and server is channel coding. In this stage, redundant information is added to source coded feature-vector data to protect against transmission errors, effectively trading bandwidth for robustness. This redundant information allows errors in transmitted data to be detected, or may even be sufficient for errors to be corrected. Data is also packetized for transmission at this stage. Since error detection and correction techniques are most effective on randomly distributed errors, it is advantageous to choose a packetization scheme that helps achieve this.

Error concealment is the third and final stage in feature vector transmission. Here, redundant information added during channel coding is utilized to detect and correct errors, much as in any digital transmission scheme. In the case of distributed speech recognition, however, we can go one step further, employing our knowledge of speech signals to handle errors. For instance, we may reconstruct feature vectors through repetition, interpolation, splicing or substitution (of which repetition usually performs best). Requesting that data be retransmitted is avoided, due to the real-time nature of speech recognition.

A number of standards have been developed for distributed speech recognition, chiefly by the Aurora working group of the European Telecommunications Standards Institute (ETSI). The first of these (ETSI ES 201 108) specifies feature-extraction processing in the front-end, source coding and channel coding (Pearce 2003a). The second (ETSI ES 202 050) upgrades this standard, adding a noise-robustness component (Pearce 2007). Both standards specify a bit rate of 4.8 kbps between client and server. These standards have also been extended (ETSI ES202 211 and ETSI ES 202 212) to include support for recognition of tone-dependent languages (Pearce 2003b; Pearce 2005).

In terms of hardware, distributed speech recognition requires sufficient resources on a mobile device to perform the front-end feature-extraction computation. In addition, since most noise robustness and reduction techniques are applied in the time domain, these must also be carried out on the device. Such processing, however, is minimal in comparison with the feature scoring and search stages of recognition, and can be carried out effectively on mobile DSPs. This hardware is often already present in modern baseband chipsets.

As for network recognition, the chief advantage of distributed speech recognition is a reduction in computational complexity on the mobile device. The chief disadvantage, again, is the power required to transmit frame vectors to the remote back-end. The bandwidth specified by the ETSI standards for these transmissions does not differ greatly from that required by standard mobile telephony codecs, in the range of 6–13 kbps. Finally, both network and distributed recognition systems are, of course, entirely dependent on network access.

2.4.4 An alternative approach: Custom hardware

While traditionally speech recognition on mobile devices has fallen into one of the three categories described in this section, an alternative approach is possible – that of custom hardware. In the following section we describe this approach.

2.5 Custom hardware for mobile speech recognition

In previous sections, we considered existing mobile hardware platforms, the hardware-resource requirements of speech recognition, and three traditional approaches to recognition on mobile devices: embedded, network and distributed systems. In the remainder of this chapter, we discuss a new approach to mobile speech recognition, that of dedicated custom hardware. In this approach, just as for the multimedia and digital television capabilities discussed in Section 2.2.1, custom hardware is added to a mobile device for performing automatic speech recognition.

2.5.1 Motivation

Network and distributed speech recognition systems allow for complex recognition tasks to be performed on mobile devices, with perhaps some impact on accuracy arising from transmission errors. Unfortunately, such systems are only suitable for mobile devices equipped to communicate with a remote server, costly in terms of power consumption and subject to the availability of wireless networks. Embedded systems, by contrast, are capable of performing standalone recognition on mobile devices, but are limited in performance by the hardware resources available.

Dedicated hardware, in the form of ASICs, offers a solution to these problems. Provided one is willing to expend the nontrivial design effort and cost for custom silicon, many of the hardware-related impediments we have previously discussed, such as limited ALUs and memory bandwidth, can be designed away. Indeed, work by Brodersen (2002) demonstrates this point in a rather dramatic fashion. We reproduce these results in Figure 2.10. The horizontal axis in this figure shows several chip designs published in the IEEE International Solid State Circuits Conference (ISSCC). The designs are grouped by architecture and are in the somewhat older CMOS technologies rather than today's technology, but the results still hold. The vertical axis measures energy efficiency in units of millions-of-operations-per-milliwatt (MOPS/mW). An 'operation' in this study is a 16-bit integer arithmetic computation, such as multiplication.

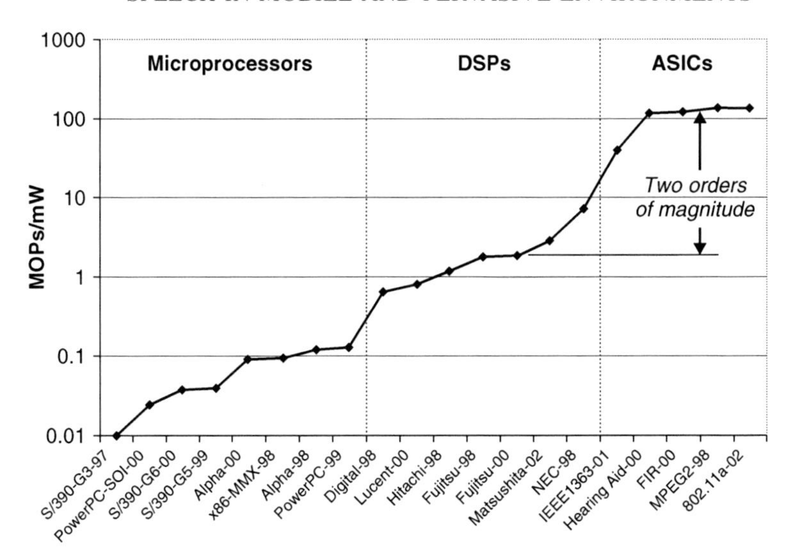

Figure 2.10 Energy efficiency (MOPS/mW) across different hardware designs (Brodersen 2002). Horizontal axis shows several ISSCC chip designs, vertical axis measures energy efficiency in MOPS/mW. Dedicated ASIC designs are dramatically more energy efficient.

We seek the largest MOPS for the fewest millwatts, so larger numbers are better here.

The plot shows that specialized DSPs outperform microprocessors by approximately two orders of magnitude, and are themselves outperformed by dedicated ASICs by another two orders of magnitude. This suggests that the cost of the flexibility of software-running processors is a dramatic loss of power-efficiency.

It is the case that speech recognition differs somewhat from the applications studied by Brodersen, in that memory operations rather than processing dominate performance. Nonetheless, preliminary results indicate that similar improvements in efficiency are possible. We now detail our own speech recognition hardware implementation work that has led us to this conclusion.

We base our hardware implementation on the Sphinx 3.0 recognizer previously described and profiled in Section 2.3. Throughout our discussion of this hardware implementation, we will assume as our workload the WSJ 5000-word recognition task. We note, however, that our hardware is capable of running both smaller tasks and considerably larger ones. Our approach to hardware will be similar to that which we took for Sphinx 3.0 itself, in that we will consider in turn hardware implementations for each of the three main Sphinx 3.0 recognition stages: feature extraction, feature scoring and search.

2.5.2 Hardware implementation: Feature extraction

We extract the same 13 MFCCs as Sphinx 3.0, performing all numerical operations using fixed-point arithmetic (Lin *et al.* 2006, 2007). In order to determine the minimum bitwidths needed to avoid affecting the overall word error rate (WER), intermediate values after each signal processing stage were profiled. As in most embedded speech recognizers, we choose fixed-point arithmetic, as the dynamic range of intermediate values is not large enough to justify the use of floating-point arithmetic and its more expensive functional units. Unlike software speech recognition, however, we are not restricted to registers that are powers of two, and can therefore reduce bitwidths to their most optimal values without restriction. This also allows us to use functional units that consume less area and power, decreasing the amount of wiring required. Minor optimizations include merging the Hamming window into the first stage of the fast Fourier transform (FFT), and reusing memory between FFT and power calculations. Final bitwidths are displayed in Figure 2.11, and the resources used in Table 2.4.

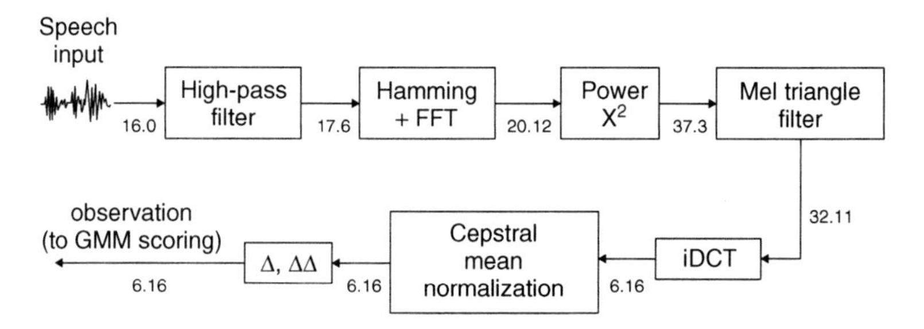

Figure 2.11 Block diagram and bitwidths of feature extraction.

Table 2.4 Resources used for feature extraction.

Gates	SRAM	Multipliers
79K	12 KBytes	7

This design requires approximately 6250 clock cycles to process one frame of speech (10 ms), so to achieve real-time recognition the feature-extraction logic needs only to run at an extremely modest 625 KHz. The design can be further optimized to increase sharing of SRAM and multipliers, which would decrease the area used but also slightly increase the clock frequency required to achieve real-time recognition. Since feature extraction can easily meet the timing constraint for real-time recognition, in the mobile domain it makes most sense to sacrifice some performance to decrease chip area.

2.5.3 Hardware implementation: Feature scoring

A straightforward evaluation of a GMM requires floating-point arithmetic involving the mixture weights, means and standard deviations, and also a complex exponential calculation (Lin *et al.* 2006; 2007). This problem is typically simplified by transforming the entire computation into the *log domain*: that is, rather than manipulating raw likelihood values we work instead with log-likelihood values. This greatly simplifies most of the computation and reduces underflow errors in the search stage, but does add the new wrinkle that we must perform addition directly in the log domain. Multiplying two numbers in the log domain is trivial: $\log(ab) = \log(a) + \log(b)$. Adding two numbers in the log-domain (which means computing $\log(a + b)$, given only $\log(a)$ and $\log(b)$) is more troublesome. Furthermore, the constant values for the log-domain analogs of the mixture weights, means and standard deviations still need to be read in from off-chip memory. Some algebraic tricks, however, can reduce this log-add computation to a large lookup table with a manageable number (about 100 000) of elements. See Lee (1988) for the mathematical details.

We also profiled the intermediate computation values for feature scoring to determine the smallest bitwidths that would not affect WER. The final bitwidths and architecture can be found in Figure 2.12, and the resources used in Table 2.5. Some important implementation details include:

- *Fixed-point and floating-point numbers.* This design uses a combination of fixed-point and customized floating-point arithmetic. During profiling, we found that the observation values did not possess a large dynamic range, making fixed-point notation the obvious choice. Since most of the 'hard' operations here are multiplications, and floating-point multipliers are smaller than fixed-point ones, we chose to convert the fixed-point intermediate results into floating-point form, and later reconvert back to fixed-point. Moreover, the GMM standard deviations exhibit a large dynamic range, which would require more bits to store using fixed-point notation than floating-point notation. At each step, our custom fixed-point and floating-point formats require less than 32 bits without affecting the overall WER.

- *Reduced bitwidths for constants.* We represent GMM means in fixed-point format with six integer bits and nine fraction bits (i.e. 6.9 format), log-normalized inverse standard deviations in floating-point format with eight bits for the mantissa and eight for the exponent, and log-normalized mixture weights using 22-bit integers. As each Gaussian mixture component must reference the 39 mean and standard deviation values before

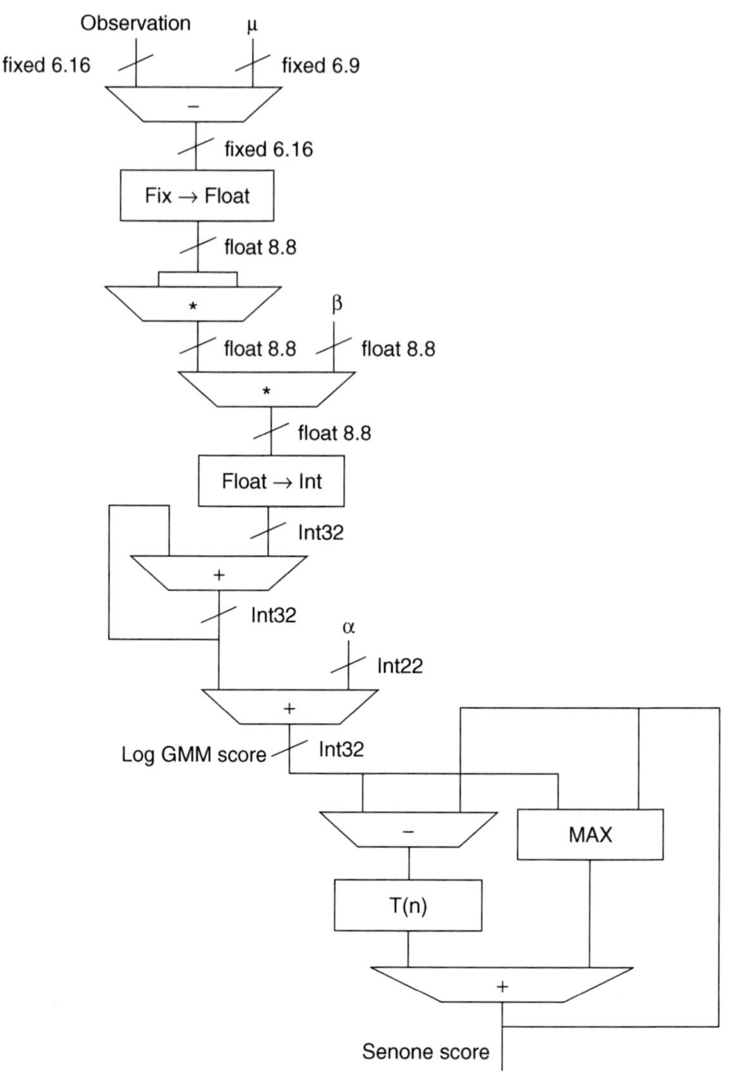

Figure 2.12 Block diagram and bitwidths of feature scoring. μ, GMM means; β, log-normalized inverse GMM standard deviations; α, log normalized GMM mixture weights.

the single mixture-weight value, we store in 32 bits one mean, one standard deviation and one bit of the mixture-weight value. In this way, the constants needed for each dimension are stored in a word with width a power of two, and by the time the mixture-weight value is needed it has been serially read in. This reduces the memory bandwidth required by

Table 2.5 Resources used for feature scoring.

Gates	SRAM	Multipliers
10K	34 KBytes	3

51%. Such bit-level arithmetic optimizations would seem horrendously complex if we were attempting to implement these in software on a conventional processor. In custom hardware, however, these aggressive bit-level optimizations are essentially free for the taking, and provide surprising boosts to overall performance.

- *Calculating logarithmic additions.* Rather than use a large on-chip lookup table to calculate the logarithmic addition, we separated the essential computation into four domains and curve-fit each domain with a third-order polynomial. Our profiling shows a lower-degree polynomial was not sufficiently accurate and reduced the recognition accuracy. Although this method takes more time to calculate than a simple lookup table, the calculation is performed only once per mixture component, hence the extra latency is hidden during the computation of these components. This change successfully replaces a very large lookup table with a single multiplier, without incurring any other cost.

Assuming sufficient DRAM bandwidth, the feature scoring logic can achieve real-time recognition at a clock speed of 129 MHz. Feature scoring is algorithmically quite simple, and so requires much less logic than feature extraction. However, since it must perform the some computation many times, it takes a great deal more execution time to complete. Since the feature-scoring unit is so small, one could easily decrease the time spent in this phase by doubling or tripling the number of feature-scoring units, provided there is enough memory bandwidth to support these additional units.

There have been previous attempts to implement feature scoring with GMMs on FPGAs (to accelerate recognition, not render it entirely in hardware), and to compare with these we also implemented our design on a Xilinx FPGA. The results are shown in Table 2.6.

Our design uses the fewest slices, block RAMs and multipliers, while providing identical functionality. This is explained by the optimization we performed. The large difference in block RAMs used is due to the fact that we do not calculate the logarithmic addition offset using a lookup table. The difference in multipliers arises because we use a reduced floating-point representation for multiplications instead of a fixed-point format.

Table 2.6 Comparison of FPGA GMM scoring implementations.

Implementation	Slices	Block RAMs	Multipliers
Marcus and Nolazco-Flores (2005)	3815	29	8
Schuster *et al.* (2006)	1527	30	6
Our design	**985**	**1**	**3**

2.5.4 Hardware implementation: Search

The final recognition stage in the Sphinx 3.0 software recognizer is HMM-based search. In this stage, feature scores from the previous stage are applied to find the most likely sequence of the HMMs Sphinx uses to model speech. This stage may be further subdivided into three phases for each frame: Viterbi scoring, transitioning between HMMs and consultation of a large trigram language model.

In order to motivate our architecture for the search stage, we note that the state of the search computation in any given frame is completely specified by two datasets: those HHMs currently being processed (the *active HMM list*), and the *word lattice* (which stores recognized words). We note that the active HMM list:

- is sufficiently large to require off-chip storage;
- requires the greatest bandwidth of any data accessed by search during recognition, yet must be read and written with equal measure;
- contains the data that drives the recognition process.

The word lattice, by comparison, is small and infrequently accessed. This suggests a hardware architecture based on the following approach:

1. Stream active HMM data on-chip.
2. Process each active HMM in turn to yield a new set of active HMMs, in much the same way a conventional processor would process instructions.
3. Stream these off-chip again.

There are two main advantages to this approach. First, since active HMM data is read sequentially, it is possible to hide much of the latency in accessing this data while also taking advantage of burst-access modes provided by memories such as DRAM. Second, since it is always possible to look ahead to unprocessed active HMMs, accesses to much of the static speech-model data needed to process an active HMM can be effectively scheduled, reducing latency.

It turns out such an approach is feasible, with one small piece of additional machinery. Of the three phases that make up the search process – Viterbi scoring, transitioning and language-model consultation – the Viterbi scoring and transitioning phases may be cast in a form suitable for operating on streamed HMMs. The language model phase, however, requires random access to active HMM data. To circumvent this, we introduce a new memory structure, the *patch list*, which accepts HMM data from the language model (Bourke 2004; Bourke and Rutenbar 2008). This data is then combined with (or *patches*) active HMMs during the scoring phase of the following frame.

The search-hardware architecture is based around three processing engines, corresponding to the three phases of the Sphinx search recognition stage. These are the scoring engine, the transition engine and the language engine, respectively. Figure 2.13 illustrates this architecture.

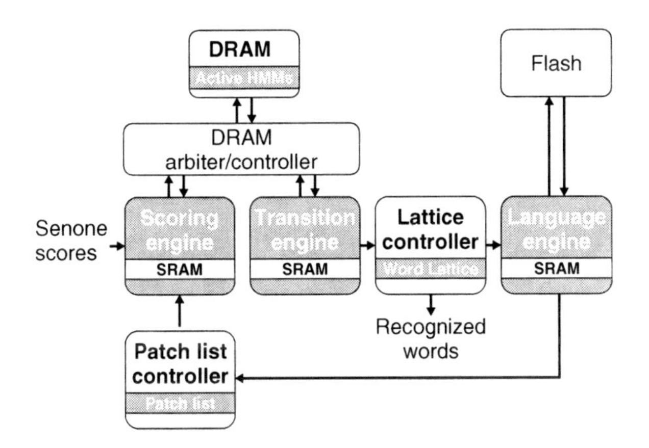

Figure 2.13 Hardware architecture for back-end search.

Communication between the three processing engines is achieved via the three *dynamic* (i.e. read *and* write) memories in the design: the active HMM list, the word lattice and the patch list. The active HMM list links the scoring and transition engines, the word lattice links the transition engine and language engine (while storing likely recognized words) and the patch list links the language engine back to the scoring engine. The active HMM list is stored in DRAM, while the word lattice (managed by the lattice controller) and patch list (managed by the patch controller) are small enough to reside in on-chip SRAM.

Recognition of a frame of speech in our hardware architecture proceeds as follows. The scoring engine first reads in active HMMs from the previous frame, calculates new state probabilities for each using the current senone scores, and writes these back to DRAM. Each newly scored HMM is then read by the transition engine, where it is either pruned, written back to DRAM alone, or written back to DRAM with any additional HMMs to which it may transition. While processing each active HMM, the transition engine notes any that complete a word, and generates a word lattice entry if necessary. These word lattice entries are then used by the language engine to determine which words should continue to be considered. Data for such words are entered into the patch list, and used to update the active HMM list during scoring of the next frame.

Thus far, we have only discussed dynamic memory in our architecture. There remains the large quantity of *static* (i.e. read-only) speech-model data required by each processing engine to perform its task. As shown in Figure 2.13, this is stored in either on-chip SRAM or external flash memory for the trigram language model (due to its size). For a standard 5000-word WSJ benchmark (Paul and Baker 1992), on-chip SRAM requirements are 11.7 Mbit, divided amongst 46 individual SRAMs – certainly a large amount, but not without precedent in today's technologies. The six largest on-chip SRAMs account for 8.8 Mbit of this total.

For the same 5000-word benchmark, the trigram language model requires 21 MByte of external flash memory. In order to efficiently locate trigrams within this memory, we employ a novel cuckoo hashing technique (Pagh and Rodler 2001). Two hash functions are used, with two bins per hash. We choose cyclic redundancy codes (CRCs) codes (Peterson and Brown 1961) as hash functions since these may be calculated extremely efficiently in hardware.

We list some important implementation details for this design in Table 2.7.

Table 2.7 Selected implementation details for hardware search.

	Scoring engine	Transition engine	Language engine
State machines	5	5	6
Pipelines	1 × 4 stage, 1 × 5 stage	1 × 2 stage	None
Register bits	4230	1384	2407
SRAMs	15	8	20
SRAM size	7.1 Mbit	960 Kbit	3 Mbit

2.5.5 Hardware implementation: Performance and power evaluation

For the implementation just described, we now discuss briefly some results indicating the potential of dedicated hardware for mobile speech recognition. In this evaluation we assume fabrication of our design in 90 nm CMOS technology.

With regard to the DRAM and flash subsystems that support our hardware, we assume relatively narrow 16-bit interfaces, and model these after commercial parts currently in production (Samsung 2005, 2006). These are mobile-specific parts with low power consumption, but support considerably less bandwidth than those designed for conventional computing platforms. In particular, maximum DRAM bandwidth is only 440 MB/s, insufficient for even feature scoring to meet real-time requirements if only a single GMM scoring unit is used. Accordingly, we configure our design with three GMM scoring units, yielding an average DRAM bandwidth requirement of 173 MB/s for feature scoring. Combined with a bandwidth requirement of 210 MB/s for the search stage, the total DRAM bandwidth for the design comes to 383 MB/s, well within the limit imposed by our DRAM subsystem.

Contention between the feature scoring and search stages for DRAM access is resolved by simply time-multiplexing this resource; the three GMM scoring units are granted exclusive access to DRAM to calculate three frames of senone scores and control then passes to the search stage to update the HMM probabilities for those frames. A small amount of additional SRAM and logic is required to support this.

We choose a clock rate of 100 MHz for our design. At this rate we are able to perform at least a single SRAM read or 32-bit integer operation within a cycle. With three GMM scoring units, this clock rate is also more than sufficient to meet real-time requirements.

In terms of accuracy, our design currently yields results identical to the Sphinx 3.0 software recognizer. For the 5000-word WSJ task we therefore maintain the same word error rate as a similarly configured Sphinx 3.0 of 6.707%.

Power consumption

We estimate the power consumption of our implementation by breaking the design into its constituent components and considering each separately. These components are the arithmetic operations, SRAM accesses, DRAM accesses, flash accesses, memory controller operations, control logic, and clock, register and pin power. We discuss each in turn.

The power consumption of individual arithmetic operations is determined by scaling values measured for similar operations in 90 nm CMOS technology (Markovic 2006); we then simply count operations. The total power consumed for arithmetic operations comes to 73.9 mW. Of this, 71.1 mW is consumed by feature extraction and scoring, in which the most complicated operation performed is 32-bit multiplication. Search consumes only 2.8 mW, the most complicated operation being 6-bit constant multiplication.

While feature scoring is considerably more complicated than search with respect to arithmetic operations, the reverse is true for memory operations. SRAM power consumption for feature extraction and scoring comes to only 0.74 mW, compared to 1.71 mW for search. Search consumes more power due to the need to perform more memory operations (dynamic power), and a requirement for vastly more on-chip SRAM (leakage power). SRAM power consumption is determined using Cacti 5.2 (Thoziyoor *et al.* 2008).

External DRAM and flash memory power is determined by considering the proportion of clock cycles in which memory is active, while also taking into account burst and sleep modes, and comes to 94 mW for DRAM and 11.3 mW for flash. We take DRAM controller power to be 5 mW (Horowitz, pers. comm.).

For our backend search (which we have not yet implemented at gate level), we assume 10 000 gates of random control logic, consuming 10 mW. Clock and register power consumption is determined using the method outlined in Markovic (2006). Performing this calculation for a 100 MHz clock and approximately 6500 total register bits yields a result of 10.1 mW. We calculate

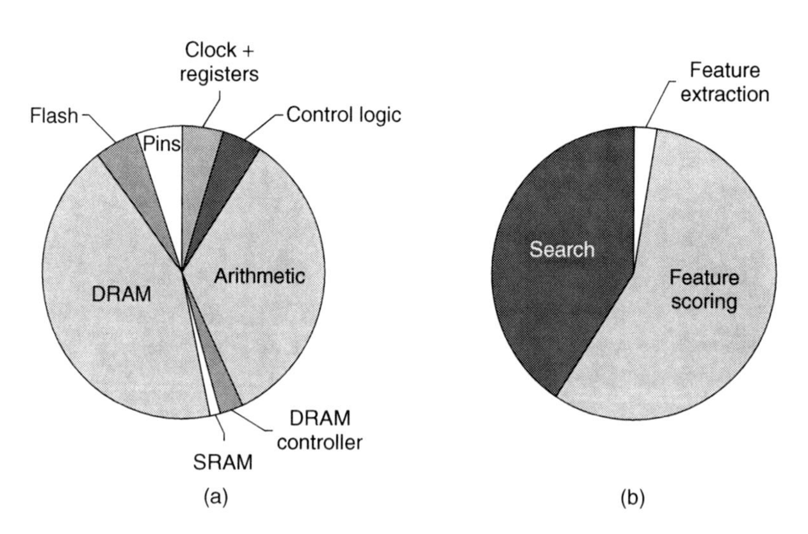

Figure 2.14 Power consumption of dedicated hardware.

pin power to be 11 mW (Horowitz, pers. comm.). The contributions of each power component to the total power consumption of our design are shown in Figure 2.14a, while in Figure 2.14b we show the power consumed by feature extraction, scoring and search.

2.5.6 Hardware implementation: Summary

For the hardware implementation described above, our initial results indicate that even at a modest clock rate of 100 MHz, it is possible to perform speech recognition for a 5000-word task in real time. Moreover, this is possible with total power consumption of approximately 220 mW, a figure close to that permitted by the extremely restrictive power budgets of modern mobile devices, and two orders of magnitude lower than conventional processors. Future work in this area can only lead to even greater performance and power improvements, and may potentially result in mobile devices that run recognition tasks of much greater complexity than would otherwise be possible.

2.6 Conclusion

Mobile platforms offer limited computational resources – a significant side effect of constraints on their cost, their size and their batteries. As a consequence, speech recognizers on mobile platforms usually make one of two unavoidable decisions: (1) to reduce the capability of a software-based recognizer hosted on the platform, or (2) to move some part of the recognition computation off the platform. In this chapter we suggested a third alternative: *render the recognizer itself in hardware on the mobile platform*. This is the path taken by graphics applications such as video playback, which are no longer handled in software, even on mobile phones. This chapter described the hardware-based solution, explained its novel constraints and opportunities, and discussed recent results targeting low-power custom silicon recognizers.

Moore's law will continue to improve the resources we have available on handsets. But just as our colleagues working on the desktop or in enterprise space, we may find that the future does not hold in store a limitless set of speed improvements for basic processors. We may soon find that multicore architectures are our essential building blocks. In enterprise applications, this means we can run several atomic recognizers in parallel on the same physical hardware. This is useful in, for example, call center telephony applications. It is less obvious how such an architecture will enhance a smartphone's speech recognition capabilities. Today's workhorse recognizers are not deeply, internally parallelized. Decomposing an atomic recognizer into a set of communicating, parallel computations is an exciting area for new research. But if we are to disassemble our favorite recognizer architectures and rebuild them for a parallel future, might we not also consider moving them away from software and into a more hardware-oriented future? We have ample data (see again Brodersen's famous plot in Figure 2.10) that custom hardware can offer many orders of magnitude improvement in power efficiency. We have the historical precedent of video playback, which started as a novelty implemented in software, and evolved to a necessity implemented in hardware on our smartphones.

Several years of recent work at Carnegie Mellon University suggest that hardware-based speech recognition has many advantages (Bourke 2004; Bourke and Rutenbar 2008; Lin and Rutenbar 2009; Lin *et al.* 2006; Lin *et al.* 2007; Yu and Rutenbar 2009). The field remains relatively new, although there are some quite recent efforts in FPGA-based recognizers (Choi *et al.* 2009; Fujinaga *et al.* 2009), and the related problem of moving speech onto new graphics processing unit (GPU) architectures (Chong *et al.* 2009). There remains a wide range of interesting questions. For example, should we be fully replacing the recognizer with hardware or just adding essential hardware assists for difficult computations like GMM scoring? Will the widespread adoption of finite-state transducer architectures (Garner 2007; Mohri *et al.* 2002) radically change the sort of architectures that have been explored for hardware speech recognition? Is hardware better deployed to improve the performance and power efficiency of the backend server clients in a network or a distributed recognizer architecture? These are all open questions, and we believe the next few years will see increasing interest in the idea of speech recognizers implemented in custom silicon.

Bibliography

Ambatipudi R. (2003) Powering next generation mobile devices. *IEEE Power Engineer*, **17**(2), 41–43. http://ieeexplore.ieee.org/Xplore/login.jsp?url=http%3A%2F%2Fieeexplore.ieee.org%2Fiel5%2F8455%2F27027%2F01200453.pdf%3Farnumber%3D1200453&authDecision=-203

ARM Holdings (2007) ARM136JF-S and ARM1136J-S technical reference manual.

ARM Holdings 2009 http://www.arm.com/products/cpus/families/cortexfamily.html. Accessed November 2009.

Baker, J. (1975) The DRAGON system – an overview. *IEEE Transactions on Acoustics, Speech, and Signal Processing*, **23**, 24–29. http://ieeexplore.ieee.org/xpl/freeabs_all.jsp?arnumber=1162650

Bocchieri, E. and Blewett, D. (2006) A decoder for LVCSR based on fixed-point arithmetic *IEEE International Conference on Acoustics, Speech and Signal Processing*, vol. 1, pp. 1113–1116. http://ieeexplore.ieee.org/xpl/freeabs_all.jsp?arnumber=1660220

Bourke, P. J. (2004) *A Queue-based Architecture for Hardware Speech Recognition*, Master's thesis, Carnegie Mellon University. http://www.ece.cmu.edu/research/publications/b/

Bourke, P. J. and Rutenbar, R. A. (2008) A low-power hardware search architecture for speech recognition. *International Conference on Spoken Language Processing*, pp. 2102–2105, Brisbane, Australia. http://www.ece.cmu.edu/~rutenbar/pdf/rutenbar-interspeech08.pdf

Brodersen, R. W. (2002) Low-voltage design for portable systems *IEEE Solid State Circuits Conference*. http://ieeexplore.ieee.org/xpl/freeabs_all.jsp?arnumber=992302

Brooks, D., Tiwari, V., and Martonosi, M. (2000) Wattch: A framework for architectural-level power analysis and optimizations, *27th International Symposium on Computer Architecture (ISCA)*, pp. 83–94. http://ieeexplore.ieee.org/xpl/freeabs_all.jsp?arnumber=854380

Brown, P. F. (1987) *The acoustic-modelling problem in automatic speech recognition*, PhD thesis, Carnegie Mellon University. http://dl.acm.org/citation.cfm?id=913535

Burd, T. D., Pering, T. A., Stratakos, A. J. and Brodersen, R. W. (2000) A dynamic voltage scaled microprocessor system. *IEEE Journal of Solid-State Circuits* **35**, 1571–1580. http://ieeexplore.ieee.org/Xplore/login.jsp?url=http%3A%2F%2Fieeexplore.ieee.org%2Fiel5%2F4%2F19075%2F00881202.pdf%3Farnumber%3D881202&authDecision=-203

Burger, D. and Austin, T. M. (1997) The SimpleScalar tool set version 2.0. *Technical Report 1342*, Computer Sciences Department, University of Wisconsin. http://www.simplescalar.com/docs/users_guide_v2.pdf

Cardinal, P., Dumouchel, P. and Boulianne, G. (2009) Using parallel architectures in speech recognition. *International Conference on Spoken Language Processing*, pp. 3039–3042. http://www.isca-speech.org/archive/interspeech_2009/i09_3039.html

Carey, D. (2007) Handset adds twists for TV viewing. *Under the hood – a supplement to EETimes*, 28–34. http://www.nxtbook.com/nxtbooks/cmp/uth051407/index.php?startid=Cover1

Chandrakasan, A. P. and Brodersen, R. W. (1992) Low-power CMOS digital design. *IEEE Journal of Solid-State Circuits*, 473–484. http://ieeexplore.ieee.org/xpl/freeabs_all.jsp?arnumber=126534

Chapparo, P., Gonzalez, J. and Gonzalez, A. (2004) Thermal-effective clustered microarchitectures. *First Workshop on Temperature-Aware Computer Systems (TACS-1)*. http://iccd.et.tudelft.nl/Proceedings/2004/22310048.pdf

Choi Yk, You, K., Choi, J and Sung, W. (2009) VLSI for 5000-word continuous speech recognition. *International Conference on Acoustics, Speech and Signal Processing*, pp. 557–560. http://ieeexplore.ieee.org/Xplore/login.jsp?url=http%3A%2F%2Fieeexplore.ieee.org%2Fiel5%2F4912736%2F4959496%2F04959644.pdf%3Farnumber%3D4959644&authDecision=-203

Chong, J., Gonina, E., Yi. Y and Keutzer, K. (2009) A fully data parallel WFST-based large vocabulary continuous speech recognition on a graphics processing unit. *International Conference on Spoken Language Processing*, pp. 1483–1486. http://www.isca-speech.org/archive/interspeech_2009/i09_1183.html

Cormie, D. (2002) The ARM11 Microarchitecture White Paper.

Davis, S. and Mermelstein, P. (1980) Comparison of parametric representations for monosyllabic word recognition in continuously spoken sentences. *IEEE Transactions on Acoustics, Speech, and Signal Processing* **28**, 357–366.

Digalakis, V., Neumeyer, L. and Perakakis, M. (1999) Quantization of cepstral parameters for speech recognition over the World Wide Web. *IEEE Journal on Selected Areas in Communications*, **17**, 82–90. http://ieeexplore.ieee.org/xpl/freeabs_all.jsp?arnumber=1163420

Franco, H., Zheng, J., Butzberger, J., Cesari, F., Frandsen, M., Arnold, J., Ramana, V., Gadde, R., Stolcke, A. and Abrash, V. (2002) DynaSpeak: SRI's scalable speech recognizer for embedded and mobile systems. *Proceedings of the Second International Conference on Human Language Technology Research*, pp. 25–30. http://citeseerx.ist.psu.edu/viewdoc/summary?doi=10.1.1.156.7028

Fujinaga, T., Miura, K., Noguchi, H., Kawaguchi, H. and Yoshimoto, M. (2009) Parallelized viterbi processor for 5,000-word large-vocabulary real-time continuous speech recognition FPGA system. *International Conference on Spoken Language Processing*, pp. 1483–1486. http://www.isca-speech.org/archive/interspeech_2009/i09_1483.html

Garner P. N. (2007) *A Weighted Finite State Transducer Tutorial*. IDIAP-Com 03-2008, IDIAP Research Institute, Martigny, Switzerland. http://publications.idiap.ch/index.php/publications/show/144

Henning J. L. (2006) SPEC CPU2006 benchmark descriptions. *ACM SIGARCH Computer Architecture News*, **34**(4), 1–17. http://dl.acm.org/citation.cfm?id=1186737

Hermansky, H. (1990) Perceptual linear predictive (PLP) analysis of speech. *Journal of the Acoustical Society of America*, **87**, 1738–1752. http://asadl.org/jasa/resource/1/jasman/v87/i4/p1738_s1?isAuthorized=no

Hetherington I. L. (2007) PocketSUMMIT: Small-footprint continuous speech recognition *International Conference on Spoken Language Processing*, pp. 1465–1468. http://www.isca-speech.org/archive/interspeech_2007/i07_1465.html

Horowitz M. A. and Dally, W. (2004) How scaling will change processor architecture. *International Solid-State Circuits Conference*, pp. 132–133. http://www.isca-speech.org/archive/interspeech_2007/i07_1465.html

Huggins-Daines, D., Kumar, M., Chan, A., Black, A., Ravishankar, M. and Rudnicky, A. (2006) PocketSphinx: a free, real-time continuous speech recognition system for hand-held devices. *IEEE International Conference on Acoustics, Speech, and Signal Processing*, pp. 185–188. http://ieeexplore.ieee.org/xpl/freeabs_all.jsp?arnumber=1659988

Itakura, F. (1975) Minimum prediction residual principle applied to speech recognition. *IEEE Transactions on Acoustics, Speech, and Signal Processing*, **23**, 67–72. http://ieeexplore.ieee.org/xpl/freeabs_all.jsp?arnumber=1162641

Jankowski Jr. C (1992) *A comparison of auditory models for automatic speech recognition*. Master's thesis, Massachusetts Institute of Technology.

Kim, H. and Cox, R. (2001) A bitstream-based front-end for wireless speech recognition on is-136 communications system. *IEEE Transactions on Speech and Audio Processing*, **9**, 558–568. http://ieeexplore.ieee.org/xpl/freeabs_all.jsp?arnumber=928920

Kuon, I. and Rose, J. (2006) Measuring the gap between FPGAs and ASICs. *Proceedings of the 2006 ACM/SIGDA 14th International Symposium on Field Programmable Gate Arrays*, pp. 21–30. http://dl.acm.org/citation.cfm?id=1117205

Kuon, I., Tessier, R. and Rose, J. (2008) *FPGA Architecture: Survey and Challenges*. Vol. 2.2 of *Foundations and Trends in Electronic Design Automation*. Now Publishers Inc. http://www.nowpublishers.com/product.aspx?doi=1000000005&product=EDA

Lee, K. (1988) *Large-vocabulary speaker-independent continuous speech recognition: The SPHINX system*. PhD thesis, Carnegie Mellon University. http://acl.ldc.upenn.edu/H/H89/H89-2038.pdf

Lilly B. T. and Paliwal K. K. (1996) Effect of speech coders on speech recognition performance. *International Conference on Spoken Language Processing*, pp. 2344–2347. http://ieeexplore.ieee.org/xpl/freeabs_all.jsp?arnumber=607278

Lin, E. C. and Rutenbar R. A. (2009) A multi-FPGA 10x-real-time high-speed search engine for a 5000-word vocabulary speech recognizer. *International Symposium on Field Programmable Gate Arrays*, pp. 83–92. http://dl.acm.org/citation.cfm?id=1508141

Lin, E. C., Yu, K., Rutenbar R. A. and Chen, T. (2006) Moving speech recognition from software to silicon: the In Silico Vox project. *International Conference on Spoken Language Processing*, pp. 2346–2349. http://citeseerx.ist.psu.edu/viewdoc/summary?doi=10.1.1.64.475

Lin, E. C., Yu, K., Rutenbar R. A. and Chen, T. (2007) A 1000-word vocabulary, speaker-independent, continuous live-mode speech recognizer implemented in a single FPGA. *International Symposium on Field Programmable Gate Arrays*, pp. 60–68. http://dl.acm.org/citation.cfm?id=1216928

Marcus, G. and Nolazco-Flores, J. (2005) An FPGA-based coprocessor for the SPHINX speech recognition system: early experiences. *International Conference on Reconfigurable Computing and FPGAs*. http://ieeexplore.ieee.org/xpl/freeabs_all.jsp?arnumber=1592509

Markovic D. M. (2006) *A Power/Area Optimal Approach to VLSI Signal Processing*. PhD thesis, University of California at Berkeley. http://techreports.lib.berkeley.edu/accessPages/EECS-2006-65.html

Masuoka, F., Asano, M., Iwahashi, H., Komuro, T. and Tanaka, S. (1984) A new flash E2PROM cell using triple polysilicon technology. *International Electron Devices Meeting*, pp. 464–467. IEEE, San Francisco. http://ieeexplore.ieee.org/Xplore/login.jsp?url=http%3A%2F%2Fieeexplore.ieee.org%2Fiel5%2F9950%2F31922%2F01484523.pdf%3Farnumber%3D1484523&authDecision=-203

Mayorga, P., Besacier, L., R. L and Serignat J. F. (2003) Audio packet loss over IP and speech recognition. *Automatic Speech Recognition and Understanding*, pp. 42–46, Virgin Islands, USA. http://ieeexplore.ieee.org/xpl/freeabs_all.jsp?arnumber=1318509

Miller, J. (2006) Reduce power consumption in the mobile handset's radio chain. *EETimes Mobile Handset DesignLine Newsletter*. http://www.eetimes.com/design/power-management-design/4016188/Reduce-power-consumption-in-the-mobile-handset-s-radio-chain

Mohri, M., Pereira, F. and Riley, M. (2002) Weighted finite-state transducers in speech recognition. *Computer Speech and Language*, **16**, 69–88. http://www.sciencedirect.com/science/article/pii/S0885230801901846

Moore G. E. (1965) Cramming more components onto integrated circuits. *Electronics Magazine*, **38**(8), 114–117. http://www.sciencedirect.com/science/article/pii/S0885230801901846

National Semiconductor (2009) www.national.com/analog/audio. Accessed November 2009.

Newman, M., Patterson, G., Roberts, M., Kamal-Saadi, M., Winterbottom, D. and McQueen, D, (2007) Mobile industry outlook 2008. Technical report, Informa PLC, London, England.

Pagh, R. and Rodler, F. (2001) Cuckoo hashing, in *Lecture Notes in Computer Science*, Springer-Verlag, pp. 121–133. http://dl.acm.org/citation.cfm?id=1006426

Parihar, N., Schlueter, R., Rybach, D. and Hansen, E. (2009) Parallel fast likelihood computation for LVCSR using mixture decomposition. *International Conference on Spoken Language Processing*, pp. 3047–3050. http://www.isca-speech.org/archive/interspeech_2009/i09_3047.html

Paul D. B. and Baker J. M. (1992) The design for the Wall Street Journal-based CSR corpus *Workshop on Speech and Natural Language*, pp. 357–362. http://dl.acm.org/citation.cfm?id=1075614

Pearce, D. (2003a) *ETSI Standard ES 201 108: Speech processing, transmission and quality aspects (STQ); distributed speech recognition; front-end feature extraction algorithm; compression algorithms, v1.1.3* European Telecommunications Standards Institute. http://www.etsi.org/deliver/etsi_es/202000_202099/202050/01.01.05_60/es_202050v010105p.pdf

Pearce, D. (2003b) *ETSI Standard ES 202 211: Speech processing, transmission and quality aspects (STQ); distributed speech recognition; extended front-end feature extraction algorithm; compression algorithms; back-end speech reconstruction algorithm, v1.1.1* European Telecommunications Standards Institute. http://www.etsi.org/deliver/etsi_es/202200_202299/202211/01.01.01_60/es_202211v010101p.pdf

Pearce, D. (2005) *ETSI Standard ES 202 212: Speech processing, transmission and quality aspects (STQ); distributed speech recognition; extended advanced front-end feature extraction algorithm; compression algorithms; back-end speech reconstruction algorithm, v1.1.2* European Telecommunications Standards Institute. http://www.etsi.org/deliver/etsi_es/202200_202299/202212/01.01.01_50/es_202212v010101m.pdf

Pearce, D. (2007) *ETSI Standard ES 202 050: Speech processing, transmission and quality aspects (STQ); distributed speech recognition; advanced front-end feature extraction algorithm; compression algorithms, v1.1.5* European Telecommunications Standards Institute. http://www.etsi.org/deliver/etsi_es/202000_202099/202050/01.01.05_60/es_202050v010105p.pdf

Pelaez-Moreno, C., Gallardo-Antolin, A. and Diaz-de Maria, F. (2001) Recognizing voice over IP: a robust front-end for speech recognition on the World Wide Web. *IEEE Transactions on Multimedia*, **3**, 209–218. http://ieeexplore.ieee.org/Xplore/login.jsp?url=http%3A%2F%2Fieeexplore.ieee.org%2Fiel5%2F6046%2F19975%2F00923820.pdf&authDecision=-203

Peterson, W. and Brown, D. (1961) Cyclic codes for error detection. *Proceedings of the IRE*, **49**, 228–235. ftp://ftp.cs.man.ac.uk/pub/apt/papers/Peterson-Brown_61.pdf

Placeway, P., Chen, S., Eskenazi, M., Jain, U., Parikh, V., Raj, B., Ravishankar, M., Rosenfeld, R., Seymore, K., Siegler, M., Stern, R. and Thayer, E. (1997) The 1996 Hub-4 Sphinx-3 system. *DARPA Speech Recognition Workshop*. Morgan Kaufmann, pp. 85–89. http://www.itl.nist.gov/iad/mig/publications/proceedings/darpa97/html/placewa1/placewa1.htm

Rabiner, L., Juang, B., Levinson, S. and Sondhi, M. (1985) Recognition of isolated digits using hidden Markov models with continuous mixture densities. *AT&T Technical Journal*, **64**, 1211–1233. http://www.citeulike.org/user/amoreno/article/5977209

Samsung (2005) K4X28163PH: 8M x16 Mobile-DDR SDRAM Datasheet, Samsung Electronics, Inc. http://www.dz863.com/datasheet-811808263-K4X28163PH_8m-X16-Mobile-ddr-Sdram/

Samsung (2006) K8C5715ETM: 16M x16 Synchronous Burst, Multi Bank NOR Flash Memory Datasheet, Samsung Electronics, Inc. http://www.dz863.com/datasheet-842324063-K8C5715ETM-FC1E_256m-Bit-16m-X16-Synch-Burst-Multi-Bank-Mlc-Nor-Flash-Memory/

Santarini, M. (2008) Eda is an ecosystem: interoperability and the future of EDA. Keynote Presentation, 21st EDA Interoperability Forum. http://www.synopsys.com/Community/Interoperability/Documents/devforum_pres/2008nov/keynote_santarini.pdf

Schuster, J., Gupta, K., Hoare, R. and Jones A. K. (2006) Speech silicon: An FPGA architecture for real-time, hidden Markov model based speech recognition. *EURASIP Journal on Embedded Systems*, **2006**, 1–19. http://jes.eurasipjournals.com/content/pdf/1687-3963-2006-048085.pdf

So, S. and Paliwal, K. K. (2006) Scalable distributed speech recognition using Gaussian mixture model-based block quantisation. *Speech Communication*, **48**, 746–758. http://citeseerx.ist.psu.edu/viewdoc/summary?doi=10.1.1.59.1281

Szepesi, T. and Shum, K. (2002) Cell phone power management requires small regulators with fast response. *EETimes Planet Analog Newsletter*. http://www.eetimes.com/electronics-news/4164128/Cell-phone-power-management-requires-small-regulators-with-fast-response

Tan, ZH, Dalsgaard, P. and Lindberg, B. (2005) Automatic speech recognition over error-prone wireless networks. *Speech Communication*, **47**, 220–242.

Tarascon J. M. and Armand, M. (2001) Issues and challenges facing rechargeable lithium batteries. *Nature*, **414**, 359–367. http://www.nature.com/nature/journal/v414/n6861/abs/414359a0.html

Texas Instruments (2009) New Multi-core OMAP 4 Applications Platform (Press release). http://www.livevideo.com/video/465365F1ABE34834953FA8E5C594B845/new-multi-core-omap-4-applica.aspx

Thoziyoor, S., Muralimanohar, N., Ahn J. H. and Jouppi N. P. (2008) Cacti 5.1. Technical Report HPL-2008-20, HP Laboratories, Palo Alto, CA. http://www.hpl.hp.com/techreports/2008/HPL-2008-20.html

Toshiba (2009) www.tmdisplay.com. Accessed November 2009.

Vasilache, M., Iso-Sipila, J. and Viikki, O. (2004) On a practical design of a low complexity speech recognition engine. *IEEE International Conference on Acoustics, Speech, and Signal Processing*, vol. 5, pp. 114–116. http://ieeexplore.ieee.org/xpl/freeabs_all.jsp?arnumber=1327060

Yu, K. and Rutenbar, R. A. (2009) Profiling large-vocabulary continuous speech recognition on embedded devices: A hardware resource sensitivity analysis. *International Conference on Spoken Language Processing*, pp. 1923–1926. http://www.isca-speech.org/archive/interspeech_2009/i09_1923.html

Zhang, Y., Parikh, D., Sankaranarayanan, K., Skadron, K. and Stan, M. (2003) HotLeakage: a temperature-aware model of subthreshold and gate leakage for architects. *Technical Report CS-2003-05*, University of Virginia. http://citeseerx.ist.psu.edu/viewdoc/summary?doi=10.1.1.7.5588

Zhu, W., Zhou, B., Prosser, C., Krbec, P. and Gao, Y. (2006) Recent advances of IBM's handheld speech translation system. *IEEE International Conference on Acoustics, Speech, and Signal Processing*, pp. 1181–1184. http://www.isca-speech.org/archive/interspeech_2006/i06_1590.html

3

Embedded automatic speech recognition and text-to-speech synthesis

Om D. Deshmukh
IBM Research, India

The focus of this chapter is to provide an introduction to the general area of automatic speech recognition (ASR) and text-to-speech (TTS) synthesis. The hope is that a thorough perusal of this chapter should equip the reader with the knowledge and confidence to dig deeper into any of the specialized topics related to ASR or TTS. In keeping with the book's focus on embedded devices, at several places the chapter makes special mention of how standard algorithms are optimized or new algorithms are devised to facilitate efficient performance of ASR and TTS techniques on embedded devices.

3.1 Automatic speech recognition

ASR, in layman's terms, refers to the process of converting spoken audio messages to text. Expectations of ASR systems are sky-high largely because of the ease with which humans can comprehend speech even in highly noisy backgrounds and with diverse accents: just think of the numerous

Speech in Mobile and Pervasive Environments, First Edition.
Nitendra Rajput and Amit A. Nanavati.

conversations that you have with your friends over the mobile riding in a crowded noisy subway with poor network coverage. A native speaker of, say, Indian English can, with very little exposure, understand American- or Chinese-accented English. Thus, it is natural to expect machines to do the same job as well as humans, if not better.

Recent advances in speech recognition and related technologies have led to highly accurate ASR systems for a variety of applications. Most of these systems achieve these high accuracies mainly by modifying one or more of their components based on the application. A generic one-size-fits-all solution is still elusive. For example, the optimal dictation ASR system has a totally different language model (LM) to the one used in an interactive voice response system. Similarly, the acoustic model (AM) in a system tailored for interactions with children would be different from the one in a system tailored for adults. Further modifications are warranted based on the computing platform, the required response time and the expected accuracy of recognition.

In this first part of this chapter, we discuss various aspects of ASR with a special focus on embedded speech recognition (ESR), which is characterized by lower computational power, lower memory speeds, lower memory capacities and lower battery capacities.

3.2 Mathematical formulation

Given a sequence of audio data, the goal of a speech recognition system is to find out the string of words that are best described by the data. This can be expressed mathematically as follows.

Let $\mathbf{O}$ be the observed sequence of audio data. The audio data is a parameterized representation of speech such that a d-dimensional vector, O_t, characterizes the salient acoustic properties of the speech signal at time t and this parameterization is done at regular time intervals (i.e. frames). $\mathbf{O}$ can then be represented as:

$$\mathbf{O} = O_1, O_2, \cdots, O_m \tag{3.1}$$

where it is assumed that the speech signal is m frames long. Also assume that the set of words $\mathbf{W}$ that can be recognized is known beforehand. $\mathbf{W}$ is referred to as the vocabulary of the ASR. Let

$$W = w_1, w_2, \cdots, w_k \tag{3.2}$$

be any string of k words. The objective of a ASR system can be stated as finding the most likely sequence of words which can occur given the acoustic data **O**. If $P(W|O)$ denotes the conditional probability of observing the word sequence W given that the audio data O was observed, then the objective of the ASR can be stated as follows:

$$\hat{W} = \underset{w}{\operatorname{argmax}} P(W|O) \tag{3.3}$$

The above equation makes several implicit assumptions (see Jelinek (1997)). The above equation can be rewritten using the Bayes' formula as:

$$P(W|O) = \frac{P(W)P(O|W)}{P(O)} \tag{3.4}$$

$P(O)$, which is the probability of observing audio data O, is independent of W and can be excluded from the maximization. $P(O|W)$ is the probability of observing audio data O when the word string W is spoken. $P(W)$ is the probability that the word string W will be spoken. The ASR objective can be restated as:

$$\hat{W} = \underset{w}{\operatorname{argmax}} P(W)P(O|W) \tag{3.5}$$

Loosely speaking, $P(W)$ is modeled by an LM, $P(O|W)$ is modeled by an AM and the detailed computation of O is captured in what is referred to as acoustic parameterization or front-end processing.

Figure 3.1 shows the various modules of the training and the testing phases of a typical ASR system.

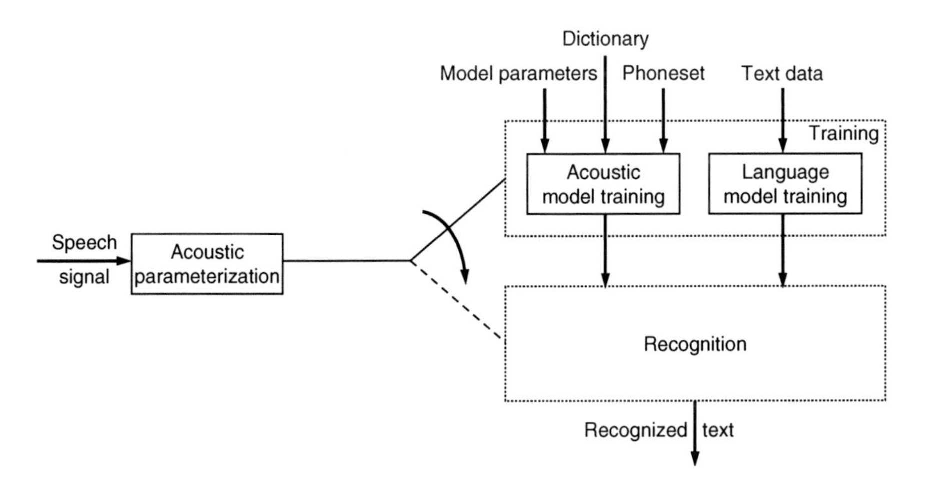

Figure 3.1 Schematic of a typical ASR system.

3.3 Acoustic parameterization

The mechanism of transforming an audio signal into a sequence of vectors that capture the audio characteristics of the speech signal is referred to as acoustic parameterization or front-end processing. Efficient and accurate parameterization of speech has led to a wide variety of feature extraction techniques and continues to attract extensive research efforts. Our understanding of human speech production and speech perception has contributed the most to this component of ASR systems. The features can be broadly categorized as: (a) uniform frame-rate features, where the same set of features is computed at regular time intervals, and (b) variable frame-rate features where different features are computed at different times based on local signal characteristics.

Figure 3.2 shows the major processing blocks in the computation of one of the popular uniform frame rate features: the mel-frequency cepstral coefficients (MFCCs). The first step in most feature computation techniques is pre-emphasis. The nature of the glottal pulse leads to lower energies in high-frequency regions of speech signals than those in mid-to-low frequencies, particularly for vocalic phones (Stevens 2000). Pre-emphasis is typically a first-order high-pass filter (of the form $1 - \alpha z^{-1}$, where α is around 0.97) that tries to restore the amplitude of high-frequency regions.

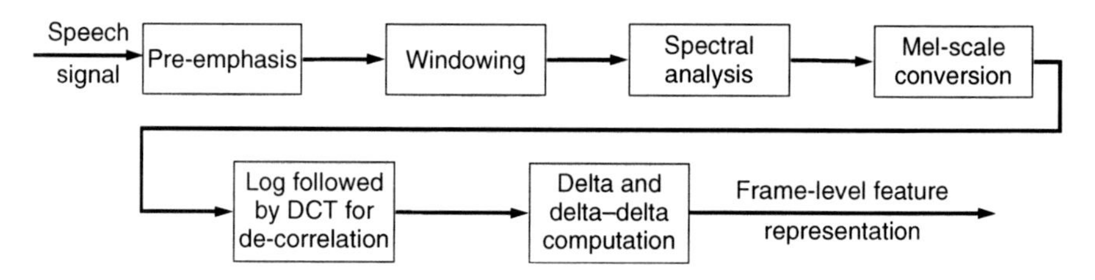

Figure 3.2 Steps in an MFCC computation.

Speech is a non-stationary signal, which means the spectral characteristics change over time. For any spectral analysis to be reasonably accurate the spectral features have to be computed over a window of the order of only tens of milliseconds (typically 25 ms). Tapering windows such as Hamming or Hanning are preferred over rectangular windows to avoid the Gibbs phenomenon (defined as the drop in accuracy of the Fourier series approximation of a function at the point of discontinuity). To reduce the loss of information due to the tapering at the edges, successive windows overlap in time, typically such that the amplitudes of the overlapping parts sum approximately to 1. The overlap also ensures temporal continuity in the spectral profile.

The spectral analysis begins by computing an N-point discrete Fourier transform (DFT) of the windowed signal. Since it is a DFT of a windowed signal, it is also referred to as a short-time Fourier transform (STFT). An N-point DFT essentially means that the speech spectrum is evaluated at N (uniformly spaced) frequency points, covering the entire spectral range of the signal. Most efficient algorithms for computation of STFT require that N be a power of 2 (Tukey 1965).

Human perception of speech and audio does not vary linearly with the frequency of the content. Several researchers have shown that human perception is more sensitive at lower frequencies than at higher frequencies (i.e. the frequency difference in two equally loud tones that humans can distinguish in the lower frequency range is much lower than that in the higher frequency range). This has led to the introduction of the mel scale (i.e. melody scale). Several formulae have been proposed to convert frequency in Hertz to its corresponding value in mel.

The most commonly used formula in ASR is (Young 2002):

$$Mel(m) = 2595 * \log_{10}\left(1 + \frac{f}{700}\right) \tag{3.6}$$

Frequency points equidistant on the mel scale are supposed to have equal resolution in human speech perception. The spectrum of a windowed speech signal on the mel scale is referred to as the mel spectrum. There are several ways to convert an STFT to a mel spectrum. One such popular method is described below. If the sampling rate of the speech signal of interest is F_s, then according to Nyquist's criterion (Oppenheim and Schafer 1999), the maximum frequency with valid content is $P = F_s/2$. The mel frequency M_P corresponding to P is computed using the above formula. The mel-scale region in the range $[0 - M_P]$ is split into B equal bins. B is the number of mel-frequency coefficients that you expect in your final feature representation. In most ASR systems B is set to 12. The centers of these B bins are the center frequencies of filters of the mel filterbank. The filters are typically triangular filters with peaks at the center frequencies and zero amplitudes at the adjacent center frequencies. The filter amplitudes for each of the filters are computed for the N frequencies used in the STFT computation. Figure 3.3 shows a template mel filterbank. The mel-spectral value at each bin is the weighted sum of the filter amplitude and the corresponding magnitude of the STFT.

The next step is to compute the log of these mel-spectral amplitudes, followed by a discrete cosine tranform. The end product is MFCCs. Note that the log of the spectrum is termed the cepstrum. The simple log operation has several advantages. Firstly, log is a highly compressing non-linearity and thus for any signal X the dynamic range of $\log(X)$ is much smaller than that of

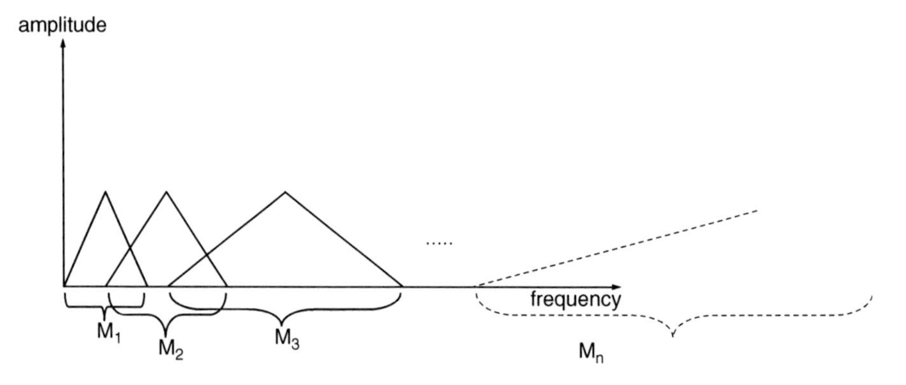

Figure 3.3 Template mel filterbank.

X. The human auditory system can respond to sounds with a wide range of amplitudes: whispers (about 15 dB) to a lion's roar (about 115 dB) and thus there is ample reason to believe that the auditory system performs certain log-like operations on the incoming audio signals. The other advantage of the log operation is that multiplication is converted to simple addition in the log domain. The effect of the different microphones used to capture the speech signal and the different channels used to transmit the signal is multiplicative in the spectral domain but is additive in the cepstral domain. A simple cepstral mean subtraction (CMS) can largely nullify the effect of variance in the microphones and transmission channels. MFCC computation in most current ASR systems is followed by a CMS procedure. A log energy feature is also added to these 12 MFCCs to make a 13-dimensional feature vector. To capture the temporal variation of these MFCCs, first- and second-order delta coefficients (also referred to as delta and delta – delta coefficients, respectively) are also computed. Delta coefficients are defined as:

$$\Delta_n = \sum_{k=1}^{K} M_{n+k} - M_{n-k} \tag{3.7}$$

where K is typically set to 2. Replacing M in the above equation with Δ leads to the delta – delta coefficients. Each frame of the speech signal is now represented by 39 MFCCs.

Many state-of-the-art ASR systems go one step further, capturing co-articulation(the effect of neighboring speech sounds on the current speech sound). This is done by concatenating the MFCC features of ± 4 neighboring frames with the features of the current frame. The $9 \times 39 = 351$-dimensional feature vector is then passed through a dimensionality-reducing technique such as linear discriminant analysis, which also leads to decorrelation of the features. (As will be seen later, decorrelation across features is also needed to keep the statistical model training tractable.)

A slightly different set of features, which have been shown to be more robust to additive and convolutive noise, are the perceptual linear prediction (PLP) (Hermansky 1990) based features. As the name suggests, these are linear predictive coefficients of a speech spectrum that has undergone some human-perception-motivated modifications. Linear prediction (LP) analysis assumes that speech signals are generated when an all-pole filter is excited by either a train of pulses (voiced sounds) or white noise (unvoiced sounds). An all-pole model implicitly assumes that the filter does not attenuate any frequencies while amplifying a certain frequencies. This is true in general for most sounds except for nasals and a few fricatives (Stevens 2000). The success of LP analysis is attributed largely to the high accuracy with which it captures the information of high-energy regions of the speech spectrum. One drawback, however, is that its treats all frequency regions with equal resolution, whereas we saw earlier that human perception is more sensitive to lower- than to higher-frequency regions. PLP computation incorporates the non-linearity of human perception by transforming the Hertz frequency axis to a perceptually motivated Bark scale. The Bark scale is very similar to the mel scale and has similar motivations as the mel scale.

PLP computation captures two other nuances of human speech perception that are not captured by MFCCs. It has been shown that human perception is more sensitive to the loudness of the middle-frequency range than the low- and high-frequency ranges. This feature is captured by multiplying the Bark spectrum by an equal-loudness curve that suppresses the low- and high-frequency regions while retaining the mid-frequency regions of the spectrum. The other aspect of human perception that PLP captures is that there is a nonlinear relationship between the perceived loudness of a sound and its intensity. This nonlinearity is adequately captured by a cube-root function. The output of the equal-loudness processing is passed through a cube-root nonlinearity to produce a 'perceptual' spectrum. An all-pole model is fitted to this perceptual spectrum to compute the PLP coefficients. The PLP features are quite often used in conjunction with RASTA processing. RASTA stands for RelAtive SpecTrA and implies passing each of the spectral components of the Bark spectrum through a bandpass filter that emphasizes spectral changes in the range of 1 to 10 Hz. The bandpass filtering is motivated by studies that have established that human hearing is relatively sensitive to variations in stimuli of the order of a few Hertz.

3.3.1 Landmark-based approach

The landmark based approach (Stevens 2002) to feature extraction and speech recognition proposes detailed analysis of speech signals only at locations that signify a major change in the acoustic characteristics of the speech signals. Moreover the type of analysis also depends on the type of change in the acoustic pattern. The approach is motivated by the proposition that speech sounds can be represented by bundles of phonetic features. The phonetic features have distinct articulatory and acoustic correlates that can be automatically extracted from the speech signal. There is also evidence that phonetic features play a crucial role in human speech perception, especially in noisy conditions. Researchers have developed acoustic parameters to automatically capture several of the phonetic features (Espy-Wilson 1994). While some effort has been made to develop ASR systems based on these variable-feature computation techniques (Juneja 2004; Salomon *et al.* 2004), there has been no large-scale use of these features in large-vocabulary continuous speech recognition. One of the main reasons for this is the lack of methods to systematically incorporate these features in the current preferred statistical backend: the hidden Markov models (HMMs). Many of the ASR systems based on variable frame-rate features use other statistical models, such as support vector machines or multi-layer perceptrons, but it is not clear how to systematically combine AMs trained using these other models with the other components of ASR such as the LM.

3.4 Acoustic modeling

3.4.1 Unit selection

We saw in the previous section how to parameterize a windowed speech signal by a vector of acoustic features. The next step is to form a coherent representation of individual speech units using these features. In almost all of the mainstream ASR systems HMMs are given the responsibility of capturing this coherent representation. The next question then is what speech unit will a HMM represent. Some of the factors to be considered are:

- naturalness;
- ease of collecting labeled data;
- number of units;
- trainability;
- context invariance.

The obvious candidates are words, phones and syllables. Each of these have been tried in one or more ASR systems with varying degrees of success.

It is easy to demarcate words in a speech signal, and modeling an entire word allows greater flexibility in capturing the numerous inter-speaker and intra-speaker variations in pronouncing the same word. One of the serious drawbacks with using the word as a unit is that the number of units increases linearly with the increase in the vocabulary of the speech signal. It becomes cumbersome to collect several repetitions of the same word from several speakers. A word-based ASR does not have the capability to recognize an unseen word. Indeed, word-units have largely been successful only in limited vocabulary tasks such as connected digit recognition.

Numbers of unique syllables will initially increase linearly with the number of unique words in the vocabulary but will saturate at high vocabulary sizes. Syllables are also good units to capture co-articulation and related pronunciation variations. The huge number of unique syllables needed to capture a large vocabulary and the lack of accurate techniques to automatically demarcate syllables in a speech signal remain the main hindrances in using syllables as ASR building blocks.

Phones, on the other hand, offer several advantages. The number of unique phones in a language is fixed, relatively low in number (typically less than 60) and does not increase as the vocabulary size increases. In a given speech signal the phone boundaries can be demarcated with relative ease and words can easily be expanded into their constitutent phones (i.e. pronunciation). This mapping of words into constituent phones is called a dictionary. Note that a word can have multiple pronunciations in a dictionary and probabilities can be assigned to each of the pronunciations, reflecting how likely the particular pronunciation is to be used. The only downside of using phones as units is that the acoustic characteristics of the phone tend to change substantially based on the identity of the phones in its left and right vicinity, the stress levels of the syllable that the phone is part of and so on. The phenomenon of phones affecting the acoustic characteristics of adjacent phones is called coarticulation and is studied extensively (for example the utterance 'did you know' is often pronounced as 'didju know'). The variability in the surface manifestation of phones based on their context implies that a single model for a phone might not be an adequate representation. To capture the coarticulatory effects, multi-phone units such as diphones and triphones have been proposed (Rosenberg *et al.* 1983; Schwartz *et al.* 1980). Assuming 50 unique phones in a language, a triphone unit can lead to $50^3 = 125\,000$ units. To adequately train models for each of these units will need prohibitively large amounts of data. The solution lies in grouping similar triphones together and training a single model for each group. Similarity across triphones can be computed in a data-driven fashion or using the knowledge of speech production. For example, all the triphones of a given center-phone with a voiced fricative (/v/, /z/) in the right context can be placed in one group while the ones with an unvoiced fricative (/f/, /s/) in the right context can be placed in a different group. If the need is felt to further reduce the number of triphones the above two groups can be combined, as the knowledge of speech production indicates that the class of voiced fricatives are the closest to the class of unvoiced fricatives in terms of their acoustic characteristics. Acoustic models that use triphone units are called context-dependent models whereas the ones that use monophone units are called context-independent models.

3.4.2 Hidden Markov models

The hidden Markov model is a powerful tool to model variable-length observations. 'Variable length' is a common phenomenon in speech signals. For example, the duration of phone /aa/ (as in 'father') in one instance can be 100 ms, whereas in a different instance by the same speaker or a different speaker it can be 150 ms. Assuming a frame rate of 10 ms, the first /aa/ is represented by 10 frames and the second one by 15 frames of observation vectors. An HMM is a collection of states that are connected to each other by transitions. Each state is characterized by two sets of probabilities.

The first of these are the state transition probabilities, which capture the probability of the current state transitioning to any of the other states. The current state i has to transition to one of the j states at every time unit (including self-transition) and thus the sum of transition probabilities for every state i has to be 1. The other set is the observation probabilities, which capture the probability of the state generating a particular feature vector. The observation probabilities can be continuous densities or discrete distributions. Continuous densities are typically parameterized as Gaussian mixture models (GMMs) and discrete distributions are typicaly reprenseted by vector-quantized codebooks (Rabiner and Schafer 1978). The Markov property states that the transition from the current state i to a future state j is independent of the exact states traversed to reach i. The 'hidden' element of an HMM is the exact state that emitted the observation: while an HMM can be associated with a frame of observation, it is almost impossible to assign the exact state that emitted the observation.

The earliest references to HMMs can be found in Baker (1979), For a detailed description of HMMs in speech recognition applications the interested reader is referred to the seminal paper by Rabiner (1989). A more formal description of HMM follows. An HMM is defined by the following parameters:

1. $S = s_t, 1, 2, \ldots, S, s_b$, the set of states in the model. State at time t is denoted by s_t. S includes an initial state S_b and a final state S_f both of

which are non-emitting. These two non-emitting states are a convenience that simplifies concatenation of HMMs of different speech units.

2. $A = a_{ij}$, the state transition matrix, where a_{ij} is the probability of transitioning from state i to state j, that is to say:

$$a_{ij} = P[s_{t+1} = j | s_t = i] \qquad \forall i, j \in S \tag{3.8}$$

Any state i at time t has to transition to one of the $j \in S$ states at time $t + 1$, which can include a self-transition to state i. Thus, the following equation has to hold for every i:

$$\sum_j a_{ij} = 1 \qquad \forall i \in S \tag{3.9}$$

In most ASR systems, only left-to-right state transitions are allowed. This can mathematically be stated as:

$$a_{ij} = 0 \qquad \text{if} \quad j < i \tag{3.10}$$

The resulting transition matrix is an upper triangular matrix and the number of parameters to learn is also reduced by $S(S - 1)/2$. This left-to-right transition also allows us to think of states as representing certain temporal portions of the speech units. For example, if phone is a unit and the HMM representing the phone has three emitting states, then the first state can be thought of as capturing the transition from the previous phone to the current phone, the second state as capturing the steady region of the phone and the third state as the transition from the current phone to the next phone.

3. $b_j(O_k)$, the probability of state j generating observation O_k. O_k is the acoustic parameterization feature vector for a given frame. It could be a 39-dimensional MFCC vector, a 26-dimensional RASTA-PLP feature or any other feature vector. It is reasonable to assume that the probability of a state j generating an observation is independent of the actual time index at which it was generated. In other words, $b_j(O_k | t = t_l) = b_j(O_k | t = t_m)$. b_j can be a continuous density or a discrete distribution. When b_j is continuous it is typically represented by a mixture of Gaussian densities. Thus, each state observation probability is modeled as a GMM. The number of Gaussian mixture components, M, is decided before hand. The density b_j can be denoted as:

$$b_j(O) = \sum_{m=1}^{M_j} w_m N(o; \mu_{jm}, \Sigma_{jm}) \tag{3.11}$$

where $N(O; \mu_{jm}, \Sigma_{jm})$ is a Gaussian PDF with mean μ_{jm} and covariance Σ_{jm}:

$$N(o; \mu_{jm}, \Sigma_{jm}) = \frac{1}{\sqrt{(2\pi)^n |\Sigma|}} \exp(-\frac{1}{2}(o - \mu_{jm})^T \Sigma^{-1}(o - \mu_{jm}) \tag{3.12}$$

The covariance matrix is assumed to be diagonal. A full covariance matrix will not only increase the computational cost but will also drastically increase the number of parameters to be learnt during model training. The feature de-correlating techniques mentioned earlier in Section 3.3 are thus necessary for efficient and accurate functioning of these models.

Figure 3.4 shows a five-state HMM with two non-emitting states and only left-to-right transitions.

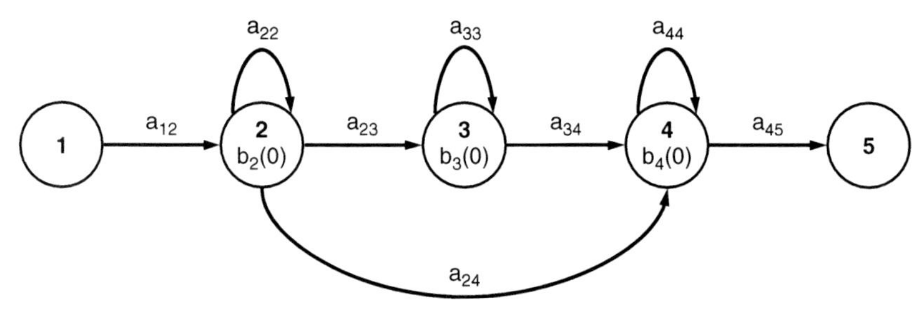

Figure 3.4 Sample five-state HMM with only left-to-right transitions. The two boundary states are non-emitting.

The HMM topology, which consists of the number of states, allowed state transitions and the form of observation emission probabilities, is fixed prior to training the model. The training procedure consists of learning the parameters of the transition probabilities and observation emission probabilities given a set of observations for each of the speech units. Several efficient recursive algorithms have been proposed for HMM model training (see for example Young (2002) for a forward – backward algorithm). The recognition task consists of computing the likelihood of a given observation sequence generated by a set of HMMs and picking the model with the maximum likelihood. One of the most efficient and widely used algorithms for this process is the Viterbi algorithm (Young 2002). The pronunciation dictionary of the ASR vocabulary and the LM restrict the number of possible HMM sequences that could have generated the given observation sequence.

Several complications arise when the phone (or triphone) HMMs are concatenated to train a continuous speech recognition system. For example, initially the individual phone models are trained in isolation using data corresponding only to that phone and independent of other phones. The concatenated HMMs should allow for a shift in the phone boundaries across training iterations, as the training objective function is the global likelihood across all the HMMs that represent a training phone sequence (i.e. word) and word-sequence (i.e. utterance) rather than each individual HMM. Many of these details and efficient solutions are covered in Jelinek (1997) and Lee (1989).

Several application-specific aspects need to be considered when building the AM. For example, if the intended use of the ASR is in classrooms then the speech data used to train the AM must be collected from children. Similarly, an AM trained on American-accented English will perform suboptimally in recognizing Indian English.

3.5 Language modeling

The next important component of an ASR system is the LM. LMs try to predict the probability of a sequence of words (or any other spoken language units) based on the structure of the language or the domain of interest. LMs are extensively used in applications outside of ASR such as machine translation, handwriting recognition, optical character recognition and others.

LMs try to estimate the probability of a word sequence $\mathbf{W} = (W_1, W_2, \ldots, W_n)$.

$$\begin{aligned} P(\mathbf{W}) &= P(W_1, W_2, \ldots, W_n) \\ &= P(W_1)P(W_2|W_1)P(W_3|W_1, W_2)\ldots P(W_n|W_1, \ldots, W_{n-1}) \\ &= \Pi_{k=1}^{n} P(W_k|W_1^{k-1}) \end{aligned} \tag{3.13}$$

where W_1^{k-1} is the sequence of $(k-1)$ words: $W_1^{k-1}) = W_1, W_2, \ldots, W_{k-1}$. Practical considerations such as data sparsity, limited storage and language structure motivate approximation of the above equation by:

$$\begin{aligned} P(W) &= \Pi_{k=1}^{n} P(W_k|W_1^{k-1}) \\ &\approx \Pi_{k=1}^{n} P(W_k|W_{k-n+1}^{k-1}) \end{aligned} \tag{3.14}$$

where n is mostly less than or equal to 4. For $n = 3$, the probability is limited to sequences of three words and such an LM is referred to as a trigram LM. Trigram LMs are the most commonly used LMs in ASR systems. The maximum likelihood estimate of a trigram is given by:

$$P(W_3|W_2, W_1) = \frac{\text{Count}(W_1, W_2, W_3)}{\text{Count}(W_1, W_2)} \tag{3.15}$$

where $\text{Count}(W_1, W_2, W_3)$ is the number of times the word sequence W_1, W_2, W_3 occurs in the training data. Similarly, $\text{Count}(W_1, W_2)$ is the number of times the word sequence W_1, W_2 occurs in the training data. This is a maximum likelihood estimate and hence by definition maximizes the likelihood of the training data with no commitments about a unseen test data (Duda *et al.* 2001). These estimates will be reasonably accurate on a test data only if the test data is similar to the training corpus. Indeed, for optimal ASR performance an LM is trained on data collected from the particular domain in which the ASR system is going to be deployed. For example, an ASR system used to transcribe a talk show on gardening will use text data related to gardening – new articles, manuals of gardening equipment and so on – to build an LM. Similarly, an ASR used to transcribe conversations between customers and call center staff of a financial institution will use data from frequently asked questions at the institute, staff training material and transcripts of previous calls to build an LM. Even in these 'in-domain' LMs there is always a concern of overfitting. For example, the probability of a trigram occuring in a test data that has never occurred in the training data is zero! This is clearly suboptimal. Moreover, Zipf's law (Manning and Schutze 1999) states that word distributions in natural languages can be best approximated by long-tail distributions and that arbitrarily large amounts of data need to be collected to estimate valid statistics on the large set of events that occur with low frequencies. To counter these concerns, all the LM building tools include some form of smoothing mechanism. One of the simplest of these is Laplace smoothing, where the count in the numerator is increased by 1 and the count in the denominator is increased by $|W|$, the vocabulary size. More sophisticated smoothing techniques include Good – Turing, Kneser – Ney and others (Chen 1998), A slightly different technique for smoothing is 'backoff' in which, if the n-gram count of a particular sequence is below a certain threshold, the counts of an $(n-1)$-gram sequence are used.

$$\begin{aligned} P(W_3|W_2, W_1) &= \frac{\text{Count}(W_1, W_2, W_3)}{\text{Count}(W_1, W_2)} \\ &\approx \frac{\text{Count}(W_2, W_3)}{\text{Count}(W_2)} \end{aligned} \tag{3.16}$$

Some of the standard LM toolkits are SRI-LM (SRI-LM 1995) and CMU-Cambridge SLM toolkit (CMU-Cambridge SLM 1997).

3.6 Modifications for embedded speech recognition

In a server-based ASR, the computational cost is not a limiting factor and hence the most accurate algorithms are used for each of the ASR modules (see Figure 3.1) irrespective of how computationally sophisticated they are. On the other hand, in embedded devices such as mobile phones, personal digital assistants or car navigation systems the computational power comes at a premium. This factor, coupled with others such as the requirement for real-time output, system size, limited battery life and heat dissipation, plays a significant role in the final design of the system. ASR systems also have to compete with other resource-intensive systems such as audio-video entertainment systems, which are present on almost all the embedded devices. There is typically no access to the OS of these devices, which means there are no avenues to implement low-level code optimization routines.

In typical servers the processor has a speed of a few gigahertz, with a capacity of tens of thousands of millions of instructions per second (MIPS), a few gigabytes of RAM and so on, whereas in a typical embedded devices (e.g. ARM processors) the processor has a speed of about 500 MHz, with a capacity of only several hundred MIPS, and a few hundred megabytes of RAM. Moreover, 64-bit processors are routine in servers whereas the processors used in embedded devices are still 16-bit or 32-bit. To put this in perspective, the LM and AM of a medium-vocabulary ASR system can each take up several hundred megabytes of storage space. The cache memory in most of the embedded devices is only of the order of a few kilobytes. In such cases, the accuracy of the ASR is traded for computationally simpler algorithms.

Each of the modules in Figure 3.1 undergoes some non-trivial modifications to perform optimally within the constraints of an embedded device. Some of these modifications are discussed in this section.

3.6.1 Feature computation

Feature computation takes only about 3–4% of the overall time taken for the entire decoding. This proportion goes down further as the vocabulary of the decoder increases. It is thus not very important to reduce the complexity of the feature computation module. However, there is a slightly different aspect that needs to be considered. The audio capturing capabilities of the microphones used in embedded devices are poor compared to closed-talking microphones. These devices are used in a wide variety of surroundings. The signal-to-noise ratio (SNR) can be as low as −10 dB and the noise conditions can be as diverse as white noise to cafeteria noise, with the added complexity that the noise environment can change abruptly. The ASR system has to include some extra processing to minimize the adverse effect of noise on the overall accuracy. This responsibility is typically shouldered by the feature computation module. The general field of improving ASR in noisy conditions is referred to as robust speech recognition and it can broadly be categorized into following three categories (Deshmukh 2006):

1. *Feature engineering.* This constitutes developing noise robust features. Truly noise robust features are still elusive and form an active area of research (Hermansky 1990). The other approach that falls in this category is feature transformation, which tries to transform the features to a domain where the effect of noise is minimal or can be easily removed. Cepstral mean subtraction (Young 2002) is one such transformation technique that is commonly used.

2. *Speech enhancement techniques.* This is where the clean speech signal is estimated from the observed noisy signal. One of the simplest and most popular of such techniques is the spectral subtraction method (Boll 1979), where the noise spectrum is estimated in non-speech regions and this spectrum is subtracted from the noisy speech signal. Several variants of this technique have been proposed and continue to be explored.

3. *The third technique tries to compensate the effect of noise by modifying the AM.* Such techniques although successful are not widely used in the ESR framework largely due to their computation-intensive nature.

Junqua (1995) covers the various challenges and approaches in robust speech recognition in great detail.

Most noise removal and/or noise-estimation techniques need long speech signals to accurately estimate the noise characteristics, whereas typical ASR interactions in mobile applications include very short utterances (e.g. short commands, the name to dial and so on). Moreover, noise estimation cannot be done in the background as the ASR system is not 'always-on' to extend the battery life of the device.

Many of the noise-reduction techniques can be approximated to reduce the computational cost while maintaining their effectiveness. One of the noise-reduction techniques that is capable of adequately handling non-stationary noise is Kalman filtering (Paliwal and Basu 1987). A Kalman filter is the optimal linear least-square error estimator of a signal corrupted by additive non-stationary noise. Kalman filtering involves estimating the noise covariance matrix at each update step. It is widely accepted that the off-diagonal elements of this covariance matrix are not as important as the diagonal elements. Often, only the diagonal elements of this matrix are estimated and updated to keep the computational cost to a minimum (Jeong *et al.* 2004).

Several modifications have been proposed to reduce the compute-load of the feature computation module also. Köhler *et al.* (2005) approximated the logarithm function (used in cepstrum computation) with a polynomial as follows: a floating-point value f is represented as:

$$f = s.m.2^x \tag{3.17}$$

where s is the sign, m is the mantissa and x is the exponent. The exponent can be thought of as an approximation of $\log_2 f$ (i.e. $\log_2 f \approx x$). The base can be changed to any number b by the multiplicative term $\log_b 2$ (i.e. $\log_b f \approx \log_b 2.x$). The error in approximation is always lower than $\log_b 2$. The approximation can be improved further by including the approximation of the logarithm of the mantissa m.

$$\log_b f \approx \log_b 2\left(\frac{1}{3}(m-1)(5-m)+x\right) \tag{3.18}$$

Köhler *et al.* (2005) show that the drop in ASR accuracy due to this approximation is only about 0.1%, while the time saved in log computation is about 80%. Some platforms, such as Intel, provide libraries for various computations optimized for their corresponding processors. Intel's Integrated Performance Primitives (Intel IPP 2011) provide efficient computation of fast Fourier transforms. Utilizing these libraries can save a substantial amount of computing time.

Most embedded platforms do not provide hardware support for floating-point operations. The alternatives are either software emulation or scaling the floating-point numbers appropriately at each stage to perform integer arithmetic (Cheng 2008). Software emualtion is computationally expensive and time-consuming. One solution is to reduce the frequency of computation or to modify the processing stages so that integer operations can be used. Converting floating-point values to integer as-is (i.e. type-casting) can lead to severe loss of resolution with adverse effects on recognition performance. A scaling factor needs to be found at each stage of floating-point operation such that the drop is resolution is minimal while integer overflow is also avoided. The scaling factor can be estimated for each step on the training data and stored for easy lookup during similar calculations on the test data. Note that the resolution obtained for a logarithm function using integer implementation is not of sufficient accuracy and the polynomial approximation mentioned earlier should be preferred.

Acoustic model training is a one-time exercise and can typically be done on the server before porting the model onto the mobile device. The computation cost of the training process itself is not critical but the cost of likelihood computation during the evaluation *is* a critical factor and needs to controlled. This cost is directly dependent on the complexity of the trained model. Thus, it is important to maintain a low level of complexity of the model. The size of the AM is also a matter of concern as the entire model is typically loaded in RAM during decoding.

It was mentioned earlier that each triphone is represented by an HMM that has N states, each of these states being modeled as a GMM with M Gaussians. One of the standard ways to reduce the model complexity is 'state-tying'. In state-tying two or more states are treated as same state. This reduces the number of parameters that need to be trained and stored. The training data corresponding to both these states are combined, which leads to a more accurately trained model. State-tying is particularly helpful for ASRs that run on resource-constrained devices, but it is also widely used in server ASRs. Several techniques exist to decide which states to tie. Some of the common techniques include the phonetic decision tree (Young *et al.* 1994) and the Bayesian information criterion (Chou and Reichl 1999). Jeong *et al.* (2004) report that state-tying alone reduced the RAM usage by about 30% while the drop in accuracy was only about 0.3%.

Another area for improvement is parameter approximation. It is widely accepted in the ASR community that performance is less sensitive to the exact variance of the Gaussians than to their mean values. In one extreme case, all the training data can be used to compute one variance for all the Gaussians across all the states. This is called the global variance (GV) approach. The GV approach leads to a drastic drop in the number of parameters to be trained and hence the footprint of the model, but also has a noticeable effect on performance. A modified global variance (MGV) approach has been developed, where the Gaussians share the same global variance vector but each Gaussian has its own weighting value with which to scale the variance vector. The drop in accuracy due to the MGV approach is shown to be minimal. A substantial drop in the AM size and a substantial improvement in the Gaussian likelihood computation can be achieved by quantization. In Köhler *et al.* (2005), the mean and covariance values are quantized into 16-bit codebooks and stored in such a way that a single 32-bit memory access fetches a mean value and the corresponding variance. Such a quantization reduces the AM size by 50%, with a negligible drop in navigation-command recognition accuracy while also increasing the decode time. Using 8-bit codebooks further reduces the AM size and the decode time but with slightly greater drop in the recognition accuracy. In this case a single 32-bit memory access can fetch two means and corresponding variances. In Jeong *et al.* (2004) and Vasilache *et al.* (2004), the input feature vector is also quantized, which reduces the entire task of likelihood computation to that of a mere table-lookup operation. For example, if the means are quantized into 2^x levels, the variances into 2^y levels and each feature into 2^z levels, then the likelihood of each feature element can take one of $2^x 2^y 2^z = 2^{x+y+z}$ values.

Replacing Gaussian densities with artificial neural networks can lead to six times fewer parameters and thus an AM with a much smaller footprint (Riis and Viiki 2000). The efficiency of this alternate model parameterization has been established only on small-vocabulary command-control applications and not on more complex applications.

3.6.2 Likelihood computation

In likelihood computation, the most obvious step is to perform the Gaussian calculations in the log domain, where the normal distribution turns into an addition operation. This is also standard practice in many of the server-based ASR systems.

Even then the Gaussian evaluation can be resource intensive. Several approaches have been proposed to reduce the number of Gaussian evaluations.

On-demand computation

In this approach (Novak 2004), Gaussians corresponding to only the few most likely HMMs are evaluated and the rest are assigned a low default likelihood value. The most likely HMMs are dictated by the restrictions imposed by the dictionary vocabulary and the LM. This approach is not scalable, i.e. when the perplexity is very high there are many phones and hence many HMM states that are equally likely.

Selection approach

In this approach, a set of 'active' Gaussians are shortlisted, given the observation vector, irrespective of the search algorithm.

- *Hierarchical labeler approach.* In the hierarchical labeler approach (Povey and Woodland 1999), the Gaussians are placed as nodes of a graph and the closeness of the Gaussians (say in terms of the Kullback Libler distance) signifies the weight on the connecting edges. For a given speech frame, the highest scoring Gaussian is found in a hierarchical fashion and all the Gaussians close to the chosen one are also evaluated.

- *Decision-tree with hyperplanes.* In a 'decision-tree with hyperplanes' approach (Padmanabhan *et al.* 1999), a decision tree is built in the feature space with each leaf node associated with a set of Gaussians. Given a input feature vector, the closest matching node from the tree is estimated and all the Gaussians in that node are evaluated.

- *Bucket box intersection.* In the bucket box intersection approach (Fritsch and Rogina 1996), the Gaussian likelihood of a given observation vector is computed only on those Gaussians that have a likelihood of greater than a pre-defined threshold τ. For a given observation vector, this set of Gaussians can be identified efficiently using space-partitioning techniques.

If the selection approach is chosen, then blocks of feature vectors can be evaluated at a time instead of serially for more efficient utilization of the cache (Novak 2004).

To further reduce the number of computations, the likelihood is computed at a lower frame rate in steady regions of the speech signal. This is called conditional frame skipping. Several efficient techniques have been proposed for detecting these steady regions.

In a modified approach, Köuler *et al.* (2005) finds similar frames by computing Euclidian distances of consecutive frames and then computes the likelihood on the mean of consecutive similar frames. None of the frames are discarded by this approach. The other avenue to reduce computational cost is to skip the state transition probabilities. This is equivalent to assuming an equal probability of transitioning to any of the valid states. As mentioned in the previous section, parameter quantization drastically reduces the cost of likelihood computation and is often used for fast decoding (Suontausta *et al.* 2000).

Several parameters contribute to the computational complexity and memory efficiency of the search module of the recognizer. It has been shown that asynchronous stack-based decoders (S. Deligne *et al.* 2002), although memory efficient, are too complex to be usable on embedded devices. The standard synchronous Viterbi decoder is preferred. The other choice is that of graphs: static versus dynamically built graphs. A static graph is a composition of the language model (L), the pronunciation dictionary (D), which maps a word to its phone sequence and the phone-to-HMM map (M), which maps the phones to corresponding HMMs. The composition is represented as: $L \circ D \circ M$. Static graphs are clearly speed-efficient as the graph construction during runtime is avoided and the use of a single graph leads to more compact search spaces. Several memory-efficient techniques (Novak and Bergl 2004) have been developed for compiling the static graph to facilitate their use on mobile devices, yet for complex AMs and large LMs dynamic search is preferred over static graphs in resource-constrained devices.

The reader should refer to Zaykovskiy (2006) for a brief survey of ESR techniques in embedded devices. For an in-depth discussion on each of the above mentioned topics on ESR, the reader is referred to Tan and Lindberg (2008) (Part III).

3.7 Applications

There are several applications where performing end-to-end ASR on embedded devices is the only viable option and then there are others that exist solely because of (or are widely popular largely because of) the ESR component. A brief summary of these is provided below.

3.7.1 Car navigation systems

One of the first industries to have benefited from the advances in speech recognition technology, especially in embedded domains, is the automotive industry. There are several systems that offer speech control of in-car navigation, information and entertainment devices. Such systems, which can communicate with various electronic and mechanical systems as well as with humans (with or without a speech interface), are referred to as telematics systems.

Early voice-driven systems had limited capabilities and would essentially only allow navigation of a pre-existing set of menu commands using voice commands. Ford's SYNC system launched in 2007 and could recognize about 100 commands. The latest version of SYNC released in 2011 can not only recognized 100 times more commands but also has 'voice learning' capabilities that learn the nuances of the driver's voice over time. Ford, in collaboration with Nuance Communications Inc., is also working on techniques to understand and interpret the intent in the user's commands, thus allowing for more natural spoken communication between the user and the ASR system. Toyota launched its Entune Multimedia Solution in January 2011. Entune uses Voicebox's conversational voice recognition and search capabilities to provide in-vehicle entertainment, navigation and information services that can be integrated with the user's mobile phone. Users can speak in natural language (instead of memorizing commands) to get directions, play music, change radio channels, make restaurant reservations, buy movie tickets and obtain other location-specific information such as real time traffic updates, weather and gas prices. The in-built TTS system can also read out movie reviews and directions. Speech interfaces for telematics systems have almost become a standard feature, at least in premium-market vehicles.

As an aside, it is important to take into consideration social norms, local culture and users' expectations while designing an system that interacts with the users via voice. For example, an early version of a voice-driven car navigation system deployed by BMW in Germany was not well-received by drivers, even though it was technically highly accurate. The main reason was that the navigation system had a female voice and German drivers were untrusting of receiving driving directions from a 'female'.

3.7.2 Smart homes

Voice-activation of household appliances has been available in the market since as early as 2002. The early systems could recognize only a few words and only if they were spoken in a particular way. There was also a training process where the user would speak the command and perform it. For example, TV remote control training would include saying the channel name and pressing the channel number a few times. In 2006, Daewoo Electronics introduced microwave ovens and washing machines that could recognize commands and then act upon them. The devices used embedded speech-recognition technology from Sensory Inc.

ESR has vastly improved since then. Moshi, a voice-controlled alarm clock manufactured by Sensory Inc., is quite accurate for a variety of users and needs no voice training. Arcturus Networks, Encore Software and Freescale have developed a module that allows a speech interface for a range of embedded devices including the ones used in smart home appliances, healthcare monitoring devices, voice-controlled vending machines and so on. The module uses the mobile version of Carnegie-Mellon University's Sphinx speech-recognition system, Pocketsphinx, and provides capabilities to configure the vocabulary and command corpus at run-time. Freescale's ColdFire embedded processors include an enhanced multiply accumulate unit, which facilitates audio processing, running the central application and managing the network connection (Bettelheim and Steele 2010).

Although ESR has tremendous potential to improve the user friendliness of various home appliances and make homes truly smart, the field is still in its nascent stages. Speech interface to control everyday appliances has great potential to improve the quality of life of an aging population and of people with motor disabilities.

3.7.3 Interactive toys

Interactive toys that can 'listen' to what the child is saying and respond appropriately are fast becoming a reality. Hallmark recently launched a set of toys that use Sensory Inc's embedded ASR systems. These 'Story Buddies' listen to key words and phrases as the child reads a story and respond by chiming or emitting appropriate responses. Mattel, another premium toy company, has an interactive robotic toy called 'Fijit Friend', which is positioned as 'every girl's best friend'. Fijit Friend[1] can recognize about 30 phrases and issue more than 100 responses including playing music and telling jokes. More such products are set to hit the market soon.

3.7.4 Smartphones

Smartphones are a common sight these days and voice applications for these mobiles are quite common, with Google leading the way. 'Voice Actions for Android' lets users interact with Android phones using their voices. It expects them to speak in a pre-defined format to activate a particular feature. For example, saying 'navigate to [address]' will bring up the directions for that particular address. 'Google Voice Search' allows users to search the Worldwide Web by talking into their phones. It uses the motion sensors of the iPhone to decide when to start recording and alerts the user by a beep. 'Google Voice' can automatically transcribe voice mails and read out received text messages.

Both Google and the Nuance system use cloud processing for all their feature computation and recognition tasks and thus do not strictly fall into the ESR category but are worth a mention here because of the array of voice capabilities they have brought to mobile phones.

CMU's PocketSphinx, a lightweight continuous speech-recognition system is being used with great success for dictation and voicemail transcription on various smartphones.

3.8 Text-to-speech synthesis

TTS synthesis is the process of automatically converting a given text utterance from a known natural language to intelligible and natural-sounding speech. Expectations from TTS systems, like ASR systems, are sky-high again because of the ease with which humans perform the simple task of reading out loud any given text in the language(s) they are familiar with. For example, humans can, with very little effort, pronounce new words correctly, know which words to emphasize, can sense the tone of the text and accordingly change the intonation to indicate 'speech-acts' (i.e. questions, exclamations and so on). The various dimensions of human spoken language acquisition and communication that are put to use to the seemingly simple task of reading a text are still being studied and new theories are being proposed to support the experimental observations (Goodman 1994).

[1] http://www.fijitfriends.com/.

TTS is fundamentally different from recorded voice-response systems. Recorded voice-response systems store recordings of each and every word that the system is likely to be asked to say. In run time, these systems merely place these individual word recordings in the right order and play the sequence. Such systems are feasible only when the list of words to be spoken is small and known in advance. TTS systems, on the other hand, have to be slightly more versatile and intelligent. They should be able to speak words that are not present in their audio recordings (if any recordings are indeed used). They should also be able to detect the correct pronunciation of a word based on its spelling (e.g. in 'cat' the letter c is pronounced as /k/ and in 'cement' it is pronounced /s/). They should also be able to resolve ambiguity. The common example used to demonstrate this is 'Dr.': in reading out an address, 'Dr.' should be expanded to the word 'drive' and in calling out names 'Dr.' should be expanded to 'doctor'.

In this section we discuss the components of a TTS system. Various techniques to implement each of these components, recent trends in this field and ways to evaluate the performance of a TTS system are also discussed.

3.9 Text to speech in a nutshell

Figure 3.5 shows a high-level block diagram of a generic TTS system. A TTS system can loosely be thought of as a combination of two modules: the front-end module and the back-end module. These two modules are sometimes also referred to as the natural language processing module and the digital speech processing module (Dutoit 1997). The front-end module takes a raw text utterance (and in case of multi-lingual TTS systems an optional 'language-ID' parameter) as input and produces a modified textual representation of the utterance along with a set of control parameters. The back-end module processes these two inputs to produce the final speech signal. The back end needs the modified textual representation in order to 'generate' the speech elements that will constitute the final speech signal. The control parameters provide information on various aspects of connecting these speech elements to improve the intelligibility and

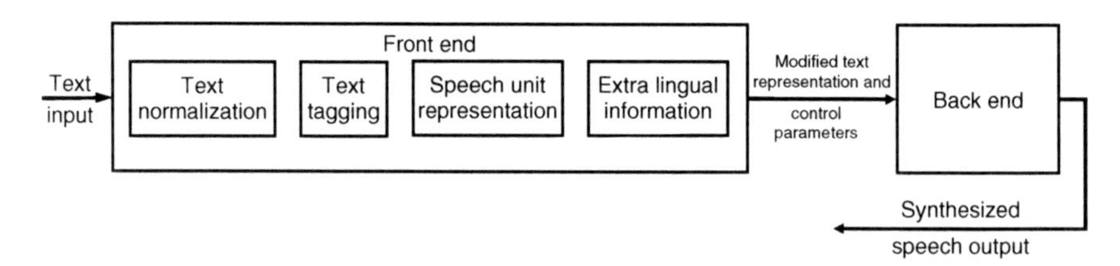

Figure 3.5 High level block diagram of a generic TTS system.

naturalness of the resulting speech signal. In a simplistic TTS, the front-end can break the input text utterance into its constituents phones and estimate the relative duration and amplitude of each. The back-end can then generate these phones, either using speech-production models or from a pre-recorded set of phones, modify them based on the relative duration and amplitude suggested by the front-end and connect them together to produce the final speech signal. Of course, in the state-of-the-art TTS systems, the front-end and the back-end modules are much more complex. Various levels of complex speech- and text-processing are included in order to extract the intended lingual and extra-lingual information that is to be conveyed by the utterance, and to pick and combine speech elements that can actually convey this information in an intelligible and natural voice.

3.10 Front end

As mentioned earlier, the front end mostly consists of the processing of the text utterance to extract the intended lingual and extra-lingual information conveyed. The front end can typically be split into four sub-modules: text normalization, text tagging, speech-unit representation and extra-lingual-information representation.

Text normalization, as the name suggests, is the process of converting the input text utterance to a standard format. The aspects of the text utterance that need special attention are punctuation, abbreviations and acronyms.

Consider the examples '2:15 am' and 'read the following: my name is Ramesh'. In the first case, the colon is to be interpreted as a delimiter of hours and minutes while in the second case the use of the colon indicates a change in sentence structure, which needs to be incorporated into the final speech output by a change in prosody. An example that is commonly used to highlight the ambiguity in expanding abbreviations is that of 'Dr', as discussed above.[2]

Acronym expansion can also be quite challenging at times. The first task is to realize that a particular word is an acronym and should not be pronounced using normal word-to-sound rules. The next task is to either look up the corresponding pronunciation of the acronym or to use a standard policy to pronounce all acronyms. For example, instances of 'I.B.M.' and 'IBM' can be standardized as 'I B M' and read out as individual letters of the English alphabet. Caution has to be exercised, as not all 'all-caps' words are acronyms. All-caps are also used as a way to highlight portions of text (e.g. emails). The reader should now appreciate, based on the above examples, that text -normalization is not a trivial process and needs to incorporate a lot of contextual information.

[2] See p. 80.

Text tagging is the process of assigning part-of-speech (PoS) tags to the words and assigning speech tags to sentences (e.g., 'question', 'anger' and so on). Accurate detection of PoS is important, as the pronunciation of a word can vary based on the PoS tag. For example, there are subtle differences in the way the word 'content' is pronounced when it is a noun versus when it is an adjective. The tense of the sentence determines the pronunciation of the word 'read' (present versus past). The PoS tag of the word 'live' leads to drastic differences in its pronunciation (/liv/ when used as a verb and /lahyv/ when used as an adverb or adjective). The PoS tagger can be a simple rule-based system Brill (1992), a data-driven technique (Brants 2000) or a hybrid approach. There are also a few standard corpora that contain word-PoS pairs.[3] The intent or the style tag of the utterance is estimated based on the temporal order of the PoS tags and the context of the utterance.

The speech-unit representation step consists of generating the phone-level pronunciations of the words of the normalized and tagged text and representing these pronunciations in terms of the speech-units that will be used by the back end. Phone-level pronunciations are generated using letter-to-sound (LTS) techniques. These techniques are also referred to as grapheme-to-phoneme techniques.

It is not always feasible (and in some cases not possible) to store the phonetic transcriptions for every word of the vocabulary, especially for morphological variations (e.g. plural, past tense and so on). LTS techniques can be broadly categorized into three types:

1. Dictionary based approach. In this approach, the focus is on storing explicitly as many pronunciations as possible. To keep the dictionary size to a minimum, pronunciation of only the elemental form a word (also called a 'morpheme') is stored and the inflectional/derivational rules required to form derived words are stored separately. The final pronunciation of the observed word is formed by combining the morpheme pronunciation with the relevant rules. The MITTALK system (Allen 1987) and the AT&T system (Levinson *et al.* 1993) use this approach.

2. Rule-based approach. This approach uses a set of LTS rules. These rules are formulated by human experts (typically linguists) who have a thorough knowledge of the particular spoken language. A separate 'exception dictionary' is maintained for words whose pronunciation does not follow any of the established rules. While this approach needs very little memory compared to the dictionary-based approach, one of the drawbacks is that is it heavily language-dependent. Dutoit (1993, Chap. 9) mentions that a combination of about 3000 rules and exceptions are

[3] http://www.cis.upenn.edu/ treebank/home.html.

enough to account for the pronunciation of 99.8% of the 60,000 words of the French Petit Robert dictionary. Note that these rules are not as simple as finding a character-to-phoneme mapping. As we said earlier, the character 'c' in 'cat' maps to the phone /k/ whereas the same character in 'cement' maps to the phone /s/.

3. Data-driven approach. This is the approach adopted by most recent TTS systems.[4] In a nutshell, these systems learn the pronunciation rules from data instead of relying on rules formulated by human experts. The LTS task can be phrased as follows: given a character sequence C, find the phone sequence Q that maximizes either the condition probability $P(Q|C)$ or the joint probability $P(Q, C)$. One of the early data-driven approaches (Black *et al.* 1998) used decision-tree algorithms to learn LTS rules from the training pairs of $words, pronunciations$. The approaches presented in Chen (2003) model the conditional as well as the joint probabilities using maximum entropy models. LTS generation has also been formulated as an HMM problem where the phones form the states and the letters are the observations generated by these states (P. Taylor 2005). In D. Wang (2011) a conditional random field model was employed with the expectation that discriminative models would lead to better global inferencing.
   Note that these LTS techniques can also be used in ASR systems to include new words in the recognizer's vocabulary.

The extra-lingual information, which implies the intent, the 'speech-act' of the spoken utterance (question, command, statement) and the emotional state of the speaker is largely captured by prosody. Variations in pitch, loudness, syllable length and inter-word pauses constitute the prosody of the spoken utterances.

Incorporating prosodic information in the TTS system can improve the naturalness of the generated speech signal. This is a two-step process. The first step extracts the prosodic information about the utterance. This information determines which words or syllables are to be grouped into a prosodic unit, the hierarchical order in which to arrange them and the intonation contour for the groups and for the entire utterance. While punctuation in the utterance helps in determining the groups, they are often not sufficient. The second step translates this information into variations of the acoustic correlates, such as pitch, loudness and duration, which will help realize the prosody. Note that there is very little margin of error, as even the slightest miscalculation can make the generated speech sound highly unnatural and even unintelligible.

[4] http://www.cstr.ed.ac.uk/projects/festival/.

The basic approaches include a separate pitch contour for different speech-acts based on the knowledge of speech production. For example, declarative sentences typically have a falling pitch while interrogative sentences have a rising pitch; all the stressed syllables have longer durations, higher pitch values and more power than the other syllables. The stressed syllables can be identified using a syllable-annotated dictionary such as CMUDict[5] or can be obtained using syllabification tools such as tsylb2.[6]

In (Hirschberg 1991), a classification and regression tree based technique is demonstrated to estimate intonational properties of an utterance based on textual analysis of its transcription. It has been shown in Syrdal and Kim (2008) that the distribution of acoustic prosodic features based on pitch, duration and power is consistent for a given speech act and varies across different speech acts. This information can be used to decide the required pitch, duration and power of the speech elements used to form the final speech output for a given utterance.

In Seimens' Papageno TTS system the prosody module has two sub-components. The first one, called the 'symbolic prosody' component, uses a multi-layer perceptron neural network to estimate the prosodic phrase boundaries and phrase accents given the input text and the corresponding PoS tags. The second component, called the 'acoustic prosody' component, uses a different neural network to estimate the pitch, duration and energy of the individual phonemes based on the output of the symbolic prosody component.

The information passed to the back end is the sequence of speech units that need to be concatenated and the control parameters that dictate the prosodic realization of these units and the overall utterance.

3.11 Back end

The back-end is largely concerned with generating the final speech signal given a simplified textual representation of the original text utterance along with a set of control parameters that dictate the interplay of the various speech units. Back-end processing can broadly be categorized into three categories based on the method employed to generate the speech signal.

3.11.1 Rule-based synthesis

In this approach, the knowledge of human speech production and speech perception is utilized to formulate rules for producing various speech sounds, either in isolation or in conjunction with each other. The perceived characteristics of a speech signal can be described as the outcome of a complex interplay of several

[5] https://cmusphinx.svn.sourceforge.net/svnroot/cmusphinx/trunk/cmudict/.
[6] http://www.itl.nist.gov/iad/mig/tools/.

parameters, which typically include formant frequencies, their bandwidths, the shape of the glottal impulses, and the spectral tilt among others. Speech sounds are modeled as the output of a linear filter that is excited by either glottal pulses (for voiced sounds such as /aa/), or noise (for unvoiced sounds such as /s/) or a combination (for sounds with simultaneous voicing and unvoicing such as /z/). The poles of the filters are the formant frequencies and their distance from the unit axis in the z-domain indicates the formant bandwidths. The values of these parameters are derived by hand and refined by analysis-by-synthesis approaches. The context information is typically captured in formant trajectories. For example, the formant trajectories from a labial unvoiced fricative such as /p/ to a vowel /aa/ would be different from the trajectories from a velar voiced fricative such as /g/ to the same vowel /aa/.

To simplify the analysis process a quasi-articulatory synthesizer such as HLsyn (Hanson and Stevens 2002) can be used. This maps higher-level quasi-articulatory parameters to the underlying acoustic parameters of the rule-based synthesizer. Use of HLsyn automatically imposes constraints on source – filter relations that occur naturally during speech production, thus preventing combinations of acoustic parameters that are impossible to achieve using human speech production apparatus.

Ruled-based synthesis systems have the advantage of providing a compact representation for a variety of voices and emotions and would seem to be the preferred choice for embedded devices. However, their major drawback is the lack of naturalness of the output speech signal and the substantial human effort it takes to build a rule-based system. It is almost impossible to capture all the acoustic effects of co-articulation and speaking styles in a compact set of rules. The generation and refinement of the rules is an iterative process, where the underlying model is first used to propose an initial set of rules. A human expert then analyzes the generated signal through auditory and visual comparisons with the corresponding natural speech signal and proposes modifications to the rules.

While Klatt and his team's pioneering work in formant snythesizers (Klatt 1980) made several seminal contributions to the area of rule-based synthesis, focus on purely rule-based synthesis systems is steadily on the decline. One of the main advantages of rule-based synthesis systems is the ability to vary the speaking rate while maintaining intelligibility and naturalness. For example, the Fonix DECtalk[7] system supports variable speaking rates from 75 to 600 words/min and occupies less than 25 KB ROM and 5 KB RAM. New voices can be added to the system with a requirement for as little as 1 KB of extra memory.

[7] http://www.fonix.com.

3.11.2 Data-driven synthesis

Data-driven methods use a minimal knowledge of human speech production. The basic approach includes recording a speech database that covers a wide range of phonetic contexts, demarcating the speech units in this database and concatenating the relevant units for a given text utterance to generate the corresponding speech signal. For this reason, they are also referred to as concatenative synthesis methods.

The very first step in this approach is to decide the basic speech-analysis unit. The two extremes of using monophones or entire words as the basic unit are not good choices. Recording monophones in isolation does not capture the co-articulation effect and concatenating the phones on demand will lead to severe audible discontinuities. On the other hand, recording entire words is impractical, not just because of the time it will take for a person to record hundreds of thousands of words, but also because of the storage requirements involved in such a design. Such an approach will also not be scalable as it cannot pronounce any word that is not in its inventory.

The most prominent choices are diphones or triphones where, unlike ASR systems, the unit begins in the middle of the steady region of the phone. This choice is dictated by the fact that, to make the generated speech signals intelligible and natural-sounding, the quality of phonetic transitions is more critical than the steady regions. A diphone unit consists of the second half (steady region to end) of the first phone and the first half (beginning to the steady region) of the second phone. A triphone unit is the same as a diphone unit, with the addition of an entire middle phone. Triphone units are better at capturing long-term co-articulation effects than diphones.

The next step is to form a diverse list of words that contains multiple instances of these units. Meaningful sentences are formed containing these words and which are recorded in quiet conditions. The units are automatically demarcated using advanced force-alignment techniques of ASR systems, with possible human interaction to minimize errors. The unit label and the corresponding locations in the database are stored in a lookup table.

The speech units can be saved in their raw form or in some parametric form. One of the biggest advantage of parameterization is the convenience of unit manipulation to meet the prosodic requirements of the target utterance. The other advantage, of course, is the drastic drop in memory requirements to save these units. This is particularly pronounced when the system is to be deployed on an embedded device.

Several speech parameterization models exist: One of the classical models is the linear prediction based model. Speech is parameterized using linear prediction coefficients (LPCs) or one of the LPC-derived representations such as line spectral pairs, which have been shown to be more robust to noise (Rabiner and Schafer 1978). In the harmonic line-spectrum approach, the complex spectrum is estimated at integer multiples of pitch for voiced frames and at the center frequencies of a DFT for unvoiced frames. The amplitude and phase spectra can be updated at regular intervals or at pitch intervals. Another popular parametric representation consists of MFCC, pitch frequency and voicing decisions.

During synthesis, the spectra of the chosen segments are estimated from the parametric representation. The spectra are then converted to time domain using inverse DFT computations and the time-domains chunks are concatenated using one of the standard overlap – add techniques.

Any of the standard compression techniques such as vector quantization can be applied to reduce the memory requirement for these parameters.

The harmonic-plus-noise model (Stylianou 2001) assumes that the speech signal is a combination of a low-frequency harmonic part and a high-frequency noise part. The speech signal, $s(t)$, can thus be represented as:

$$s(t) = s_h(t) + s_n(t)$$

$$s_h(t) = \sum_{k=-L(t)}^{L(t)} A_k(t) e^{jk\omega_o(t)t}$$

$$s_n(t) = e(t)[h(\tau, t) * b(t)]$$

where $s_h(t)$ is the harmonic part and $s_n(t)$ is the noise part. $\omega_o(t)$ is the pitch frequency and $L(t)$ is the number of pitch harmonics to be included in the harmonic part. This depends on the maximum voiced frequency ($F_m(t)$). $F_m(t)$ can be estimated by a peak-picking alogrithm. $A_k(t)$ is the complex amplitude value. The noise part, $s_n(t)$, is modeled by the output of white noise $b(t)$ filtered by a time-varying autoregressive filter $h(\tau, t)$. For every analysis, only a handful of parameters need to be estimated and saved. During synthesis, the parameters of the selected speech segment are used to generate the frame waveform. The frame-level waveforms are combined using an overlap-and-add process. Prosodic variations performed on the harmonic-plus-noise model representation approach lead to less distortions than other standard methods such as TD-PSOLA (see below).

One of the most popular overlap and add method is the time domain pitch synchronous overlap and add (TD-PSOLA) method (Moulines and Charpentier 1990). The original speech signal is windowed into overlapping segments using a window that is about 2–4 times the instantaneous value of the pitch and always centered at a pitch epoch. The pitch and duration of the windowed segments are modified (if needed) as explained below. The modified windowed segments are then overlap-added to form the reconstructed signal. Tapering windows, such as Hanning windows, ensure that the contribution of each sample in the overlapped regions is close to 1. The pitch frequency is modified by adjusting the time intervals between consecutive pitch epochs and the duration is modified by either dropping or repeating segments. Note that modification of pitch frequency indirectly modifies the duration also. The computational simplicity and the ease of modifying pitch and duration related prosodic features has contributed to the popularity of the TD-PSOLA method and its variants in the speech-synthesis field.

At the synthesis stage, speech units extracted from different contexts are concatenated to generate the required output. Several levels of smoothing techniques are applied to reduce the perceived discontinuity at the concatenated edges. Each segment is processed so that the energies at the beginning and the ending of all the segments are similar. After the concatenation, various global smoothing operations are performed, which attempt to normalize spectral amplitude across the units.

The quality of the synthesized speech is quite sensitive to these smoothing operations. Moreover, signal-processing techniques to modify the prosody of selected speech units so as to meet that of the required target can adequately change the temporal pattern of pitch and energy but are still incapable of adequately varying the spectral characteristics to capture the difference in voice quality and phonation due to prosodic changes. This reduces the naturalness of the generated speech signals even though they have high intelligibility.

This has led to the evolution of the traditional concatenative synthesis approach into the unit selection concatenative approach (Santen *et al.* 1997, Chap. 22). The unit selection approach involves a much larger database, with many more instances of the speech units of interest, than in the traditional concatenative database. The most relevant instance of the required unit is chosen by considering various aspects such as phonetic, prosodic and context relevance with the target. The basic premise is that the naturalness of the generated speech signal is much better if the speech units are concatenated 'as is' rather than when they are passed through a signal processing block to meet the characteristics of the target. A large database containing multiple instances of the same unit but with different prosodic characteristics and different contexts is more likely to contain an instance of the unit that is appropriate for the target without any modifications.

The labels for such databases have to be rich, in that not only do they have to contain information about the phone identity and their time demarcations but also about their prosodic characteristics. As mentioned earlier, phone identity and time demarcations can be obtained by force-aligned recognition. Phone-level prosodic labels can be obtained as follows: for every instance of every phone the standard prosodic parameters such as pitch frequency, amplitude, spectral energy and spectral tilt are calculated at a fixed frame rate and averaged over the duration of the phone. These parameters are normalized by computing the means and variances across all the instances of the particular phone. Prosodic labels are generated by computing the relative differences in these parameters over the preceding and following N phones. The sign and the magnitude of these differences capture the prosodic information of the phone (see section above). In embedded devices, where memory is at a premium, instances that have a similar context and similar prosodic labels can be deleted and only a few representative instances can be stored.

The next step is to select the speech units that match the required target. This can be achieved in several ways. In minimum distortion criteria (Iwahashi *et al.* 1992), the speech units that minimize a global distortion value are selected. Assume that there are S segments in the target utterance for which the 'most appropriate' inventory units have to be detected. Let u_i denote the ith inventory unit and t_j be the jth target segment. Each inventory unit and target segment is represented by an F-dimensional feature vector: $u_i = [u_{i1}, u_{i2}, \ldots, u_{iF}]$. Two types of distortion are defined: unit distortion $(I_D(t_i, u_j))$ is the distance between the inventory unit u_j and target segment t_i. Continuity distortion $(C_D(u_{j-1}, u_j))$ is the distance between the unit selected for the jth target and the $(j-1)$th target segment. The relative significance of these two distortions can be varied by multiplicative weights W_I and W_c. Also note that the contribution of an individual feature dimension for either of the two distortion measures can be varied by a $F \times F$ matrix M_f (i.e. replacing u_i by $M_f * u_i$). The cost function to be minimized can now be stated as:

$$\sum_{i=1}^{S} = W_c \times C_D(u_i, u_{i-1}) + W_I \times \sum_{j \in (p_1, p_2, \ldots, p_i)} I_D(t_i, u_j) \tag{3.19}$$

where $(p_1, p_2, \ldots, p_i)$ are the possible units that match target t_i. In Hunt and Black (1996), the speech database is represented as a state transition network where each individual unit is a separate state. The state occupancy cost is the unit distortion (I_D) and the state transition cost is the continuity distortion (C_D). Given a sequence of target segments, the task is to find the state transitions with minimum cost. A standard Viterbi algorithm can be used to find this optimal state sequence. Other methods based on decision-tree clustering (Wang *et al.* 1993) and HMMs (Donovan and Woodland 1999) have also been successfully used for optimal unit selection.

3.11.3 Statistical parameteric speech synthesis

In this approach, parameters of a statistical model are learnt during the training phase in order to represent speech signals. During the synthesis phase, appropriate model parameters are chosen to generate the required speech signal. The most commonly used statistical model is the HMM and so this approach is also referred to as the HMM synthesis approach. Figure 3.6 shows the different steps in the HMM synthesis process.

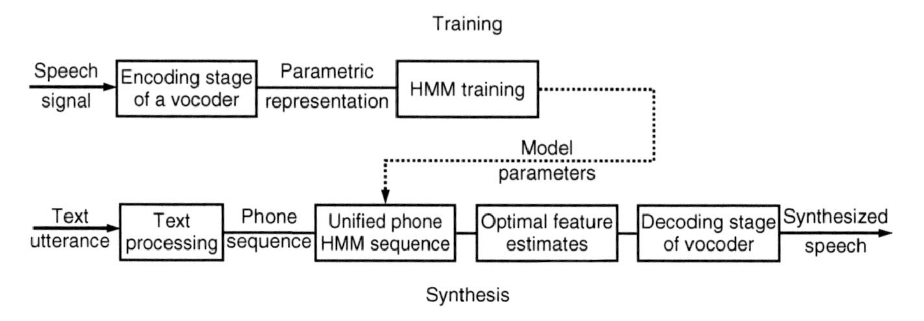

Figure 3.6 Schematic of the HMM synthesis process.

Most of the motivation and algorithmic details for the HMM synthesis method have come from technical advances in and successful implementations of HMM systems for ASR. There are a few fundamental differences though:

- The feature set contains a few extra features to capture the pitch and the excitation characteristics along with the traditional MFCC (or some variant) feature set.
- The duration density of each state is explicitly modeled, unlike in ASR where the self-transition probability implicitly captures the state duration density. This modification infringes upon the Markov assumption but the model is still referred to as HMM.
- A wider context is used to train a phone HMM and it is not restricted to just the two adjacent phones as in ASR.
- There is also an extra step of estimating parameters that will optimally generate the speech output as desired by the input text utterance.

The HMM model is sandwiched between the encoding and the decoding phases of a vocoder. The choice of the vocoder is based on factors such as the set of parameters the coder uses, the sensitivity of these parameters to noise and the quality of vocoder-constructed speech signal. Some example vocoders used are the mel log spectral approximation filter (Masuko *et al.* 1996) and a combination of a glottal inverse filter and a vocal tract filter (Raitio *et al.* 2011).

Given the orthographic transcription and the prosodic labels of the training data, the parameters of the HMMs are trained in a manner similar to HMM training for ASR systems. The prosodic labels for the training data can either be assigned by humans or can be generated by the front-end module of the TTS system (se Figure 3.5). To reduce the number of independent model parameters to be learnt and to be able to generate sounds from unseen contexts, similar phone models or states are tied. The model-tying is similar to the model-tying used for ASR systems.

In the synthesis stage, the front-end provides the phone sequence along with the prosodic labels to the HMM system. The corresponding phone HMMs are concatenated to form a unified HMM. The speech parameters are generated from this unified HMM such that the characteristics of the required speech signal are optimally represented. The details of the speech-parameter generation algorithm can be found in Tokuda *et al.* (1995) and King (2010). These parameters are then passed to the decoding stage of the vocoder, which generates the final speech signal.

One of the main advantages of HMM-synthesis techniques over concatenative synthesis techniques is that the speech signals used in HMM synthesis need not be of as high quality because the signals are not directly copy-pasted to form the synthesized signal, as happens in concatenative methods.

3.12 Embedded text-to-speech

As mentioned earlier, a rule-based synthesizer needs the least amount of memory and computation and should be a clear winner in TTS requirements on embedded devices. For example, Fonix DECtalk, one of the leading rule-based commercial synthesizers, occupies less than 25 KB ROM and 5 KB RAM. New voices can be added to the system with as little as 1 KB of extra memory required.

If the available memory is in the range of 1 MB then several concatenative and statistical parameteric synthesis techniques can be ported to embedded devices. In concatenative synthesis techniques the required memory is largely decided by the size of the audio recordings database. It is almost standard to downsample all the recordings to 8 kHz. Lower sampling rates help in saving both memory and computational cost, as the number of frames of feature computation and processing is reduced. Hoffmann (2006) shows that MPEG compression of the raw audio data leads to a compression ratio of about 14.8:1, with very little drop in intelligibility of the synthesized speech signal. Speech signals can also be stored in parameteric form using one of the several speech-coding standards. The coding, which can be compute intensive, can be done off-line on servers. The decoding algorithms, which need to be run on the embedded device, are not very resource intensive. If the embedded device is a mobile phone, then the same codec as the one used by the mobile platform can be used to encode the audio data.

The other avenue to save memory is the word-to-phone dictionary. Replacing dictionaries that contain phone expansion of most of the words with a rule-based dictionary can drastically reduce the memory requirement. Hoffmann *et al.* (2003) show that a 960-KB word dictionary can be replaced by a 62-KB rule-based dictionary, with a very small drop in the overall perceptual quality of the output signal and an almost negligible increase in the extra processing required.

HMM synthesis methods are gaining popularity in the embedded domain too, as they do not have to save any audio data. They have also been aided by technical advances in HMM-based ASR systems for the embedded devices. Advances in the ASR domain are also helping other aspects of embedded TTS systems. For example, it has been shown that a male TTS voice can easily be converted to a female voice using a knowledge of vocal tract length normalization (VTLN), which is extensively studied in the ASR domain (Eichner *et al.* 2004). Such a technique can provide a choice of TTS voices, with a minimal increase in the memory footprint.

3.13 Evaluation

There are several techniques and evaluation metrics to evaluate the performance of each of the modules of a TTS system. There are several standard evaluation techniques.[8] In this section we only cover the evaluation metrics used to evaluate the end-to-end TTS system. A TTS-generated speech signal is generally evaluated in three dimensions:

- accuracy: how many words did the system get right, not just in terms of correctness of pronunciation but also in terms of interpretation (e.g. 'Dr.' being interpreted as 'drive' or 'doctor');
- intelligibility, which quantifies the proportion of generated speech that the human listener was able to understand;
- naturalness, which estimates how likely it is for a human listener to mistake the generated speech signal for a human recorded utterance.

[8] http://www.elda.org/en/proj/tcstar-wp4/,http://festvox.org/blizzard/index.html.

There is no one universal measure that can capture the overall quality of a TTS system and can be used to compare qualities of different systems. Some of the commonly used metrics are briefly described below:

1. *Comprehension tests.* A paragraph is played to the user and he/she is asked to answer a set of questions related to the paragraph. The correctness of user's answers is an indication of the intelligibility of the synthesized paragraph.

2. *Diagnostic rhyme test (DRT) and modified rhyme test (MRT).* DRT and MRT evaluate the identifiability of individual phones. In these tests, the user is asked to choose the word he/she heard from a pair of words. The pair of words differ either in their word-initial consonant (DRT) (e.g. sit versus fit, cat versus pat and so on) or in their word-final consonant (MRT) (e.g. mob versus mop, back versus bat and so on). Similar tests include the cluster identification test and the standard segmental test.

3. *Semantically unpredictable sentence test.* In this test, users are played sentences that do not follow the semantic structure imposed by the language and thus recognizing words in one part of the sentence need not necessarily help in predicting the rest of the sentence. An example sentence would be: 'chicken roots sink in soon'.

4. *Mean opinion score* (MOS): MOS is one of the most widely used techniques to evaluate the quality of speech. Users are played a set of sentences and asked to rate the quality of each sentence on a scale of 1–5, where 1 is the poorest quality and 5 is the best. A simple average of the scores of these sentences across different users is the MOS of the system. MOS evaluation is very subjective and the outcome depends heavily on test conditions, the complexity of sentences and the quantity and quality of the users.

5. *Preference tests.* In these tests, the user is played two versions of the same sentence: one generated by the TTS system and the other a human recording or possibly generated by a different TTS system. The user is asked to note his/her preference. Collating user preferences over a large set of sentences and across a wide range of users can help rank one TTS system over the other or to compare a TTS system with human recordings.

6. *Objective perceptual tests.* As opposed to all the above tests, which need human evaluation, the perceptual evalaution of speech quality[9] and

[9] http://www.itu.int/rec/T-REC-P.862/en.

perceptual speech quality measure are objective evaluations, which do not need human intervention but exhibit high correlations with subjective human evaluations. The speech signals are processed through a series of spectro-temporal warpings to mimic the established models of human speech perception and are then compared with the reference signal to generate a MOS-like score.

3.14 Summary

Various topics in the general area of ASR and TTS synthesis were presented in this chapter, so as to enable the reader to dig deeper into any of the specific topics of his/her interest. Both ASR and TTS have a rich history incorporating more than three decades of research and engineering effort. The latest systems have come a long way from their original predecessors. There is a lot of scope for improvement, so as to bridge the gap between human – human and human – machine spoken communication. Advances in the fields of speech processing, machine learning and related areas have made it possible to incorporate a large amount of human – machine spoken communication into embedded devices. This area continues to fascinate and challenge engineers and researchers from academia and industry alike.

Bibliography

Allen, J. (1987) *From Text to Speech: The MITTALK System* Cambridge University Press, Cambridge.

Baker, J. M. (1975) The Dragon system: an overview. *IEEE Transactions on Acoustics, Speech and Signal Processing*, 23, 24–29.

Bettelheim, R. and Steele, D. (2010) White Paper: *Speech/Command Recognition*. http://www.arcturusnetworks.com/whitepapers/Speech_Recognition_WPv0-0-8a.pdf.

Black, A., Lenzo, K. and Pagel, V. (1998) Issues in Building General Letter to Sound Rules. In *Third ESCA Workshop on Speech Synthesis*, International Speech Communication Association (ISCA), 77–80.

Boll, S. (1979) Suppression of Acoustic Noise in Speech Using Spectral Subtraction. In *IEEE Transactions on Acoustics Speech and Signal Processing*, ASSP-27(2), 113–120.

Brants, T. (2000) TnT: A Statistical Part-of-Speech Tagger. In *Proceedings of the Sixth Conference on Applied Natural Language Processing*, 224–231.

Brill, E. (1992) A Simple Rule-based Part of Speech Tagger. In *Proceedings of the Third Conference on Applied Natural Language Processing*, 112–116.

Chen, S. F. and Goodman, J. (1998) *An Empirical Study of Smoothing Techniques for Language Modeling*. Technical Report, TR-10-98, Computer Science Group, Harvard University.

Chen, S. F. (2003) Conditional and joint models for grapheme-to-phoneme conversion. In *Proceedings of Interspeech 2003*, 2033–2036.

Cheng, O. (2008) Embedded Speech Recognition Systems. https://researchspace.auckland.ac.nz/handle/2292/3279.

Chou, W. and Reichl, W. (1999) Decision tree state tying based on penalized Bayesian information criterion. In *Proceedings of IEEE International Conference on Acoustics, Speech, and Signal Processing*, 345–348.

CMU-Cambridge (1997) Statistical Language Modeling Toolkit. http://mi.eng.cam.ac.uk/~prc14/toolkit.html.

Deligne, S., Dharanipragada, S., Gopinath, R., Maison, B., Olsen, P. and Printz, H. (2002) A robust high accuracy speech recognition system for mobile applications. *IEEE Transactions on Speech and Audio Processing*, 10, 551–561.

Deshmukh, Om. D. (2006) *Synergy of Acoustic-Phonetics and Auditory Modeling towards Robust Speech Recognition*. Ph.D. Thesis, University of Maryland, College Park.

Donovan, R. and Woodland, P. (1999) A hidden Markov model based trainable speech synthesizer. *Computer Speech and Language*, 13, 1–19.

Duda, R., Hart, P. and Stork, D. (2001) *Pattern Recognition*. Wiley-Interscience.

Dutoit, T. (1997) *An Introduction to Text-to-Speech Synthesis*. Kluwer Academic Publishers.

Dutoit, T. (1993) *High Quality Text-to-Speech Synthesis of the French Language*. Ph.D. Thesis, Faculté Polytechnique de Mons.

Eichner, M., Wolff, M. and Hoffmann, R. (2004) Voice characteristics conversion for TTS using reverse VTLN. In *Proc. of IEEE International Conference on Acoustics, Speech, and Signal Processing*, 1, 17–20.

Espy-Wilson, C. (1994) A feature-based semivowel recognition system. *Journal of Acoustical Society of America*, 96, 75–72.

Fritsch, J. and Rogina, I. (1996) The bucket box intersection algorithm for fast approximate evaluation of diagonal mixture Gaussians. In *Proceedings of IEEE International Conference on Acoustics, Speech, and Signal Processing*, 837–840.

Goodman, J. (1994) *The Development of Speech Perception*. MIT Press.

Hanson, H. and Stevens, K. (2002) A quasiarticulatory approach to controlling acoustic source parameters in a Klatt-type formant synthesizer using HLsyn. *Journal of Acoustical Society of America*, 112, 1158–1182.

Hermansky, H. (1990) Perceptual linear predictive (PLP) analysis of speech. *Journal of Acoustical Society of America*, 87, 1738–1752.

Hirschberg, J. (1991) Using text analysis to predict intonation boundaries. In *Proceedings of Eurospeech-91*, 1275–1278, 1991.

Hoffmann, R. (2006) Speech synthesis on the way to embedded systems. In *Proceedings of SPECOM-2006*, 17–26, 2006.

Hoffmann, R., Jokisch, O., Hirschfeld, D., Strecha, G., Kruschke, H., Kordon, U. and Koloska, U. (2003) A multilingual TTS system with less than 1 Mbyte footprint for embedded applications. In *Proc. of IEEE International Conference on Acoustics, Speech, and Signal Processing, 2003,* 1, 532–535.

Hunt, A. and Black, A. (1996) Unit selection in a concatenative speech synthesis system using a large speech database. In *Proc. of IEEE International Conference on Acoustics, Speech, and Signal Processing, 1996*, 2, 373–376.

Intel (2011) Integrated Performance Primitives 7.0 http://software.intel.com/en-us/articles/intel-ipp/.

Iwahashi, N., Kaiki, N. and Sagisaka, Y. (1992) Concatenative speech synthesis by minimum distortion criteria. In *Proceedings of IEEE International Conference on Acoustics, Speech, and Signal Processing, 1992*, 2, 65–68.

Jelinek, F. (1997) *Statistical Methods for Speech Recognition*. The MIT Press

Jeong, S., Han, I., Jon, E. and Kim, J. (2004) Memory and computation reduction for embedded ASR systems. In *Interspeech 2004*, 2325–2328.

Junqua, J-C and J-P Haton, J-P. (1995) *Robustness in Automatic Speech Recognition: Fundamentals and Applications*. Kluwer.

Juneja, A. (2004) *Speech Recognition Based on Phonetic Features and Acoustic Landmarks*. Ph.D. Thesis, University of Maryland, College Park.

King, S. (2010) *A Beginners Guide to Statistical Parametric Speech Synthesis*. Draft available at: http://www.cstr.ed.ac.uk/downloads/publications/2010/king_hmm_tutorial.pdf

Klatt, D. (1980) Software for a cascade/parallel formant synthesizer. *Journal of Acoustical Society of America*, 67, 13–33.

Köhler, T. W., Christian Fügen, C., Stüker, S. and Waibel, A. (2005) Rapid porting of ASR-systems to mobile devices. In *Proceedings of Interspeech 2005*, 233–236.

Lee, K. (1989) *Automatic Speech Recognition: The Development of the SPHINX System*. Kluwer.

Levinson, S., Olive, J. and Tschirgi, J. (1993) Speech synthesis in telecommunications. *IEEE Communications Magazine*, Nov, 31(11), 46–53.

Manning, C. and Schutze, H. (1999) *Foundations of Statistical Natural Language Processing*. The MIT Press.

Masuko, T., Tokuda, K., Kobayashi, T. and Imai, S. (1996) Speech synthesis using HMMs with dynamic features. In *Proceedings of International Conference on Acoustics, Speech and Signal Processing*, 191–194.

Moulines, E. and Charpentier, F. (1990) Pitch-synchronous waveform processing techniques for text-to-speech synthesis using diphones. *Speech Communication*, 9, 453–467.

Novak, M. (2004) Towards large vocabulary ASR on embedded platforms. In *Interspeech 2004*, 2309–2312.

Novak, M. and Bergl, V. (2004) Memory efficient decoding graph compilation with wide cross-word acoustic context. In *Proceedings of Interspeech 2004*, 281–284.

Oppenheim, A., Schafer, R. and Buck, J. (1999) *Digital Processing Speech Signals*. Prentice Hall.

Padmanabhan, M., Bahl, L. and Nahamoo, D. (1999) Partioning the feature space of a classifier with linear hyperplanes. *IEEE Transactions on Speech and Audio Processing*, SAP-7, 282–288.

Paliwal, K. K. and Basu, A.. (1987) A speech enhancement method based on Kalman filtering. In *International Conference on Acoustics, Speech, and Signal Processing*, 177–180.

Povey, D. and Woodland, P. (1999) Frame discriminant training of HMMs for large vocabulary speech recognition. In *International Conference on Acoustics, Speech, and Signal Processing*, 333–336.

Raitio, T., Suni, A., Pulakka, H., Vainio, and M., Alku, P. (2011) Utilizing glottal source pulse library for generating improved excitation signal for HMM-based speech synthesis. In *Proceedings of IEEE International Conference on Acoustics, Speech, and Signal Processing*, 2011, 4564–4567.

Rabiner, L. (1989) A tutorial on hidden Markov models and selected applications in speech recognition. In *Proceedings of the IEEE*, 77, 257–286.

Rabiner, L. and Juang, B.. (1993) *Fundamentals of Speech Recognition*. Prentice Hall.

Rabiner, L. and Schafer, R. (1978) *Digital Processing Speech Signals*. Prentice Hall.

Riis, S. K. and Viikki, O. (2000) Low complexity speaker independent command word recognition in car environments. In *Proceedings of IEEE International Conference on Acoustics, Speech, and Signal Processing, 2000*, 3, 1743–1746.

Rosenberg, A. E., Rabiner, L. R., Wilpon J. G. and Kahn D. (1983) Demisyllable-based isolated word recognition system, *IEEE Transactions, on Acoustics Speech and Signal Processing*, ASSP-31, 713–726.

Salomon, A. and Espy-Wilson, C. and Deshmukh, Om. D. (2004) Detection of speech landmarks: use of temporal information. *Journal of Acoustical Society of America*, 115, 1296–1305.

Santen, J., Sproat, R., Olive, J. and Hirschberg, J. (1997) *Progress in Speech Synthesis*. Springer-Verlag.

Schwartz, R., Klovstad, J., Makhoul, J. and Sorensen, J. (1980) A preliminary deisgn of phonetic vocader based on diphone model. In *Proceedings of IEEE International Conference on Acoustics, Speech and Signal Processing*, 32–35.

SRILM (1995) The SRI Language Modeling Toolkit http://www-speech.sri .com/projects/srilm/.

Stevens, K. N. (2000) *Acoustic Phonetics*. The MIT Press.

Stevens, K. N. (2002) Toward a model for lexical access based on acoustic landmarks and distinctive features. Journal of Acoustical Society of America, vol. 111 (4), 1872–1891, 2002.

Stylianou, Y. (2001) Applying the harmonic plus noise model in concatenative speech synthesis. *IEEE Transactions on Speech and Audio Processing*, 9, 21–29.

Suontausta, J., Hakkinen, J. and Viikki, O. (2000) Fast decoding in large vocabulary name dialing. In *Proceedings of IEEE International Conference on Acoustics, Speech, and Signal Processing, 2000*, 3, 1535–1538.

Syrdal, A. and Kim, Y-J. (2008) Dialog speech acts and prosody: considerations for TTS. In *Proceedings of Speech Prosody*, 661–665.

Tan. Z. H. and Lindberg, B. (Eds.) (2008) *Automatic Speech Recognition on Mobile Devices and over Communication Networks*. Springer-Verlag, London

Taylor, P. (2005) Hidden Markov models for grapheme to phoneme conversion. In *Proceedings of Interspeech 2005*, 1973–1976.

Tokuda, K., Masuko, T., Yamada, T., Kobayashi. T. and Imai, S. (1995) An algorithm for speech parameter generation from continuous mixture HMMs with dynamic features. In *Proceedings of Eurospeech-95*, 757–760.

Tukey. J. W. (1965) An algorithm for the machine calculation of complex Fourier series. *Mathematics of Computation*, 19, 297–301.

Vasilache, M., Iso-Sipilä, J. and Viiki, O. (2004) On a practical design of a low complexity speech recognition engine. In *Proc. of IEEE International Conference on Acoustics, Speech, and Signal Processing, 2004*, 5, 113–116.

Wang, D. and King, S. (2011) Letter-to-sound pronunciation prediction using conditional random fields IEEE Signal Processing Letters, vol. 18(2), 122–125, 2011

Wang, W., Campbell, W., Iwahashi, N. and Sagisaka, Y. (1993) Tree-based unit selection for english speech synthesis. In *Proceedings of International Conference on Acoustics, Speech and Signal Processing*, 191–194.

Young, S. (2002) The HMM Toolkit (HTK) Book http://htk.eng.cam.ac.uk.

Young, S., Odell, J. and Woodland, P. (1994) Tree-based state tying for high accuracy acoustic modelling. In *Proceedings of the Workshop on Human Language Technology*, 307–312.

Zaykovskiy, D. (2006) Survey of the Speech Recognition Techniques for Mobile Devices. *Eleventh International Conference on Speech and Computer* (SPECOM), St. Petersburg (Russia), 88–93.

4

Distributed speech recognition

Nitendra Rajput and Amit A. Nanavati
IBM Research, India

Speech processing is a computation-intensive activity. Historically, it was always done on the server. When people call up over a (landline) phone, the voice goes over the wire (channel) to the server, where the speech is processed for recognition. The transmission of the signal usually leads to a deterioration in the quality of the signal received at the server, and hence a lower quality of recognition.

With more and more mobile phones having more and more processing power, speech processing on the client is becoming a viable option. One option is to have client-only processing in which most of speech processing is done at the client side, the results then being transmitted to the server, provided the client has sufficient computing resources for the task. This was the focus of Chapter 3, on embedded speech recognition.

Another option is to have part of the processing on the client and part on the server, an approach known as *distributed* speech recognition (DSR). In this model, front-end processing is done on the client, the speech features are transmitted from the client to the server and finally the processing of speech decoding and language understanding are performed at the server side (Zhang *et al.* 2000). The idea is to have some limited processing done at the client, and send the relevant features to the server, thus getting rid of the deterioration that happens due to signal transmission and reconstruction.

Speech in Mobile and Pervasive Environments, First Edition.
Nitendra Rajput and Amit A. Nanavati.

In DSR, the focus is on using data instead of voice networks. One reason is that mobile voice networks degrade performance due to their low bit-rate speech coding and channel transmission errors. Using an 'error-protected data channel instead, recognition performance can be kept at high levels.

In this chapter, we will give an overview of DSR: how the processing is shared between the client (front end) and the server (back end), the standards activity in this space (this is crucial since there is a plethora of devices in the market), including the data transfer issues. We also relate DSR with ESR and NSR (network speech recognition). Also, for thin clients, battery power is a precious resource – we touch upon this briefly.

4.1 Elements of distributed speech processing

In the client–server DSR system architecture, the ASR (automatic speech recognition) processing is split into client-based front-end feature extraction and server-based back-end recognition, with data transmitted between the two parts via heterogeneous networks (Tan *et al.* 2005).

The value of DSR is that it provides substantial recognition performance advantages over conventional mobile voice channels, where both the codec compression and channel errors degrade performance. A codec is a compression–decompression program, which converts analog audio signals into digital signals for transmission or storage. A receiving device then converts the digital signals back to analog using an audio decompressor, for playback.

DSR also enables new mobile multimodal interfaces by allowing the features to be sent simultaneously with other information on a single mobile data channel such as GPRS. As summarized in Pearce (2009), the main benefits of DSR are as follows.

- There is improved recognition performance over wireless channels. The use of DSR minimizes the impact of speech codec and channel errors that reduce the performance from recognizers accessed over digital mobile speech channels.

- There is easy integration of combined speech and data applications. Many new mobile multimodal applications are envisioned, such as the use of speech to access wireless Internet content. The use of DSR enables these to operate over a single wireless data transport rather than having separate speech and data channels.

- There is ubiquitous access with guaranteed recognition performance levels. DSR promises a guaranteed level of recognition performance over every network. It uses the same front end and there is no channel distortion coming from the speech codec and transmission errors.

In order to reduce transmission error degradation, client-driven recovery and server-based concealment techniques are applied within DSR systems along with error-detection techniques. Server-based error-concealment (EC) techniques exploit the redundancy in the transmitted signal and this may be used independent of, or in combination with, client-based techniques.

4.2 Front-end processing

The DSR idea is based on decoupling the front-end mechanism from the recognition process. In this way the front-end mechanism can be located on a client device (such as a mobile phone) while the recognition process takes place at the server (Manolis 2001).

Figure 4.1 shows the block diagram of a DSR system. The client front-end block displays the functions performed at the client. This includes feature extraction, source coding and channel coding.

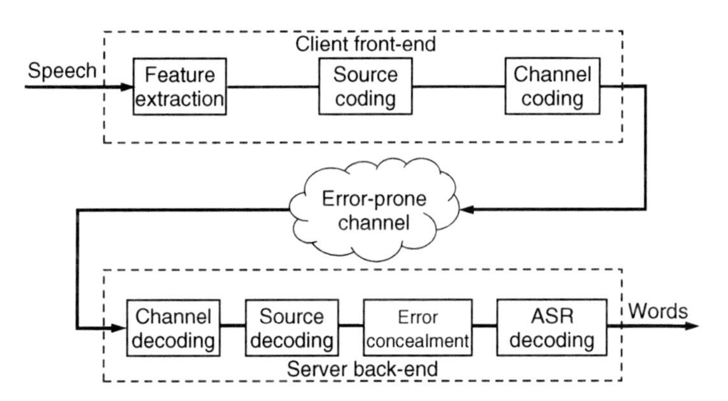

Figure 4.1 Block diagram of the DSR system (Tan et al. *2005).*

Feature extraction is the process of obtaining different features such as power, pitch and vocal tract configuration from the speech signal. Broadly, the feature extraction techniques are classified as temporal analysis and spectral analysis techniques. In temporal analysis the speech waveform itself is used for analysis. In spectral analysis a spectral representation of the speech signal is used for analysis.

The goal of source coding is to compress information for transmission over bandwidth-limited channels. One common class of coding schemes for DSR applies vector quantization (VQ) to ASR features (see Figure 4.2). Split VQ together with scalar quantization is used to compress Mel-frequency cepstral coefficients (MFCCs).

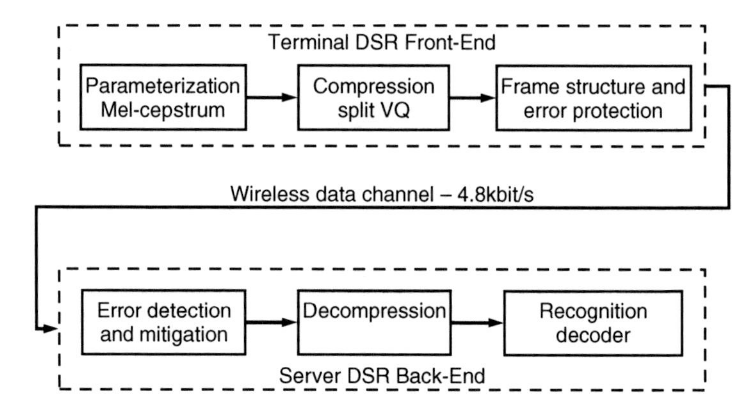

Figure 4.2 Block diagram of DSR at the next level (Pearce 2009).

While source coding aims at compressing information, channel-coding techniques attempt to protect (detect and/or correct) information from distortions (Bossert 2000). Channel coding is an error-control technique used for reliable data delivery across error-prone channels by means of adding redundancy to the data (Sklar *et al.* 2004).

Figure 4.2 describes the block diagram at another level in terms of the activities that each block performs. The following steps happen at the front end:

- the speech signal is sampled and parameterized using the mel-cepstrum algorithm
- this generates 12 cepstral coefficients along with C_0 and $\log E$
- it is compressed to obtain a lower data rate (4.8 kbps) for transmission
- the compressed parameters are formatted into a defined bitstream
- it is then transmitted over wireless/wireline transmission link to a remote server
- the parameters are then checked for transmission errors
- the front-end parameters are decompressed to reconstruct the DSR mel-cepstrum features
- these are then passed to the recognition decoder sitting on central server.

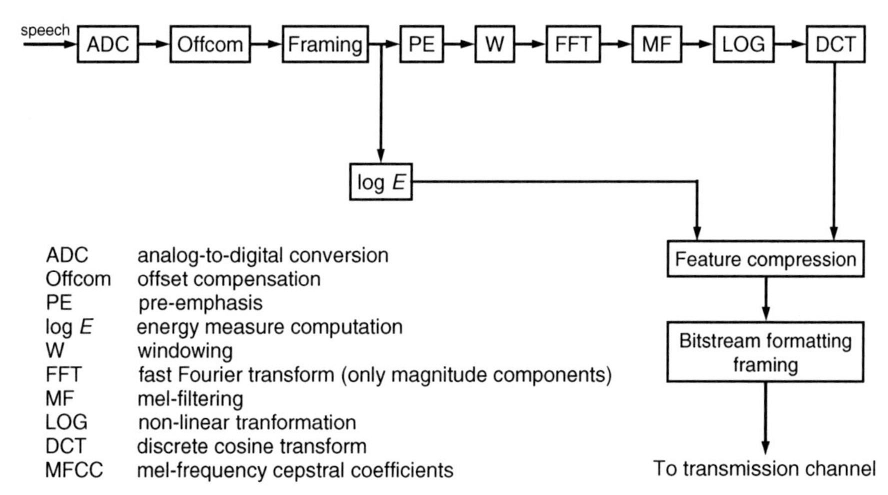

Figure 4.3 Block diagram of mel-cepstrum DSR front end standard (Pearce 2009).

Figure 4.3 shows the specification of the mel-cepstrum DSR front-end standard. The feature vector consists of 12 cepstral coefficients (C_1–C_{12}), which, together with C_0 and the log E (log energy) parameter, make up a total of 14 components. C_0 is included to support algorithms that might need it in the back end (such as noise adaption). A split VQ algorithm is used to obtain a final total data rate of 4800 bits per second (see Figure 4.2) of speech. A codebook of size 64 is used for each pair of cepstral coefficients from C_1 to C_{12} and 256 vectors are used for C_0 and energy. This results in 44 bits per speech frame. Since the parameters may be transmitted over error-prone channels, error detection bits (4 bits of cyclic redundancy check (CRC) for each pair of speech frames) are applied to the compressed data. For transmission and decoding the compressed speech frames are grouped into multiframes corresponding to 240 ms of speech. The format is such that the bits corresponding to two frames may be transmitted as soon as they are ready. This results in only 10 ms additional latency at the terminal.

4.2.1 Device requirements

In DSR, the client side has three main processing tasks: (a) capture the speech signal and process it to remove ambient noise, (b) process the enhanced speech signal to compute the acoustic features, (c) code these acoustic features for efficient transmission to the server. The first part typically involves matrix multiplications while the second part involves some basic signal processing such as FFT computation along with other mathematical operations such as matrix multiplication and logarithm evaluations. The last step is not computationally intensive. Thus, the client-side processor should either have capabilities for efficient computation of FFT (some of the advanced mobile processors from Intel have such capabilities) or the operating system should provide low-level support to optimize these operations. Most mobile processors do not provide hardware support for floating-point operations. This implies that the software should have the capability to emulate these floating-point operations.

4.2.2 Transmission issues in DSR

The deployment of ASR in networks requires specific attention due to a number of factors, such as the more complicated architecture, the limited resources in the terminals, the bandwidth constraints and the transmission errors. These network-linked issues have focused the research in network-based ASR on front-end processing for remote speech recognition, on source coding and on channel coding and EC.

Studies have shown superior performance of DSR compared to codec-based ASR. Although DSR eliminates the degradations originating from the speech compression algorithms, the transmission of data across networks still brings in a number of problems to speech recognition technology, in particular transmission errors. Figure 4.4 shows the client- and server-side efforts to handle transmission errors.

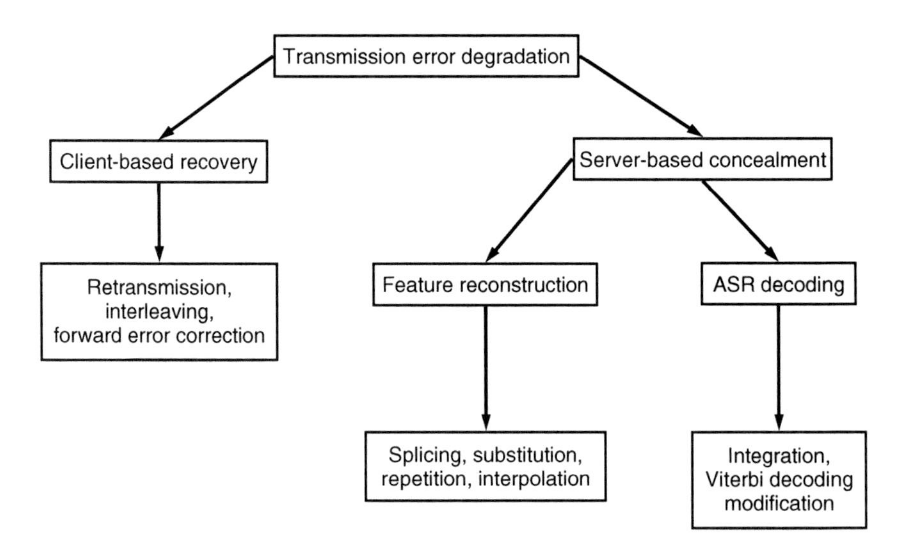

Figure 4.4 Transmission errors in DSR (Tan et al. *2005).*

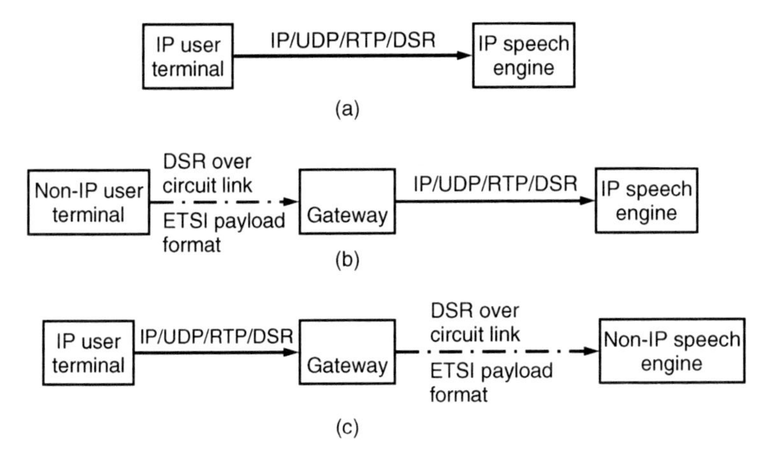

Figure 4.5 Scenarios using DSR payload format (RFC 3557 2003).

4.2.3 Back-end processing

Lossless error recovery is required in data transmission, as even a single-bit error may cause the entire data block to be discarded (Tan *et al.* 2005). In contrast, a certain number of distortions in speech features can be tolerated by the ASR decoder. This fact makes EC a feasible method to complement client-based error recovery techniques so as to mitigate the effect of the remaining transmission errors without the request for a retransmission. EC generally takes advantage of the strong temporal correlation in speech features and utilizes statistical information about speech.

The aim of EC is in general to create a substitution for a lost or erroneous packet that is as close to the original as possible. This type of concealment technique is termed as feature-reconstruction EC. EC may be conducted during the recognition decoding process as well, something which is unique to DSR. Specifically, the ASR-decoder may be modified to handle degradations introduced by transmission errors, a process termed ASR-decoder EC.

The algorithm for error mitigation consists of two stages (Pearce 2009):

- detection of speech frames received with errors;
- substitution of parameters when errors are detected.

To detect the speech frames received with errors, the four error-detection bits on each pair of frames are used first. Since errors may be missed due to overloading of the CRC, a heuristic algorithm that looks at the consistency of the parameters in the decoded frames is also used. It measures the difference between cepstral coefficients for adjacent frames and flags them as erroneous if it is greater than expected for speech. The thresholds used are based on measurements of error free speech. If this algorithm was to run continuously then the number of misfirings would be too high, therefore it is only applied in the presence of CRC errors.

4.3 ETSI standards

The work of the European Telecommunications Standards Institute (ETSI) is focused on standardizing the front-end mechanism so that all the front-end information produced by different clients is identical, no matter what kind of device the client is.

To enable widespread application of DSR in the market place, a standard for the front end is needed to ensure compatibility between the terminal and the remote recognizer. The Aurora DSR Working Group within ETSI has developed this standard (Pearce 2009).

To allow the optimization of the details of the feature extraction algorithm a reference database and experimental framework was established. The database is based on the original TIdigits database, with controlled filtering and simulated noise addition over a range of signal-to-noise ratios from 20 dB to 5 dB. A reference recognizer configuration using Entropics HTK HMM software was agreed, to allow investigation of changes in the front end. This database has been made publicly available via the European Language Resources Association. Extensive experimentation has been performed using this and other internal databases to agree on the feature extraction and to test the compression algorithms. The first DSR standard was published by ETSI in February 2000.

Mel-cepstrum was chosen as the first standard because of its widespread use throughout the speech recognition industry. It was acknowledged, however, that a front end that had better performance in dealing with background noise would be desirable.

The ETSI DSR basic front end defines the feature-extraction processing together with an encoding scheme, (ES 201 108 2000). The decoding algorithm at the server conducts two calculations to determine whether or not a frame pair is received with errors, namely a CRC test and a data consistency test. The ETSI standard uses a repetition scheme in its EC processing to replace erroneous vectors.

The features sent to the server-based recognition system are commonly cepstral coefficients, so the speech-enhancement techniques must be applied at the client side. ETSI included noise-robustness techniques in the advanced front-end standard (ES 202 050 2007).

4.3.1 Basic front-end standard ES 201 108

This ETSI standard (ES 201 108 2000) presents the standard for a front end to ensure compatibility between the terminal and the remote recognizer. Mel-cepstrum is used as the front end. The standard specifies algorithms for front-end feature extraction and their transmission, which together form part of a system for distributed speech recognition. The specification covers the following components:

- the algorithm for front-end feature extraction to create mel-cepstrum parameters;
- the algorithm to compress these features to provide a lower data transmission rate;
- the formatting of these features with error protection into a bitstream for transmission;
- the decoding of the bitstream to generate the front-end features at a receiver together with the associated algorithms for channel error mitigation.

This standard also describes the DSR front-end algorithm based on the mel-cepstral feature-extraction technique. The specification covers the computation of feature vectors from speech waveforms sampled at different rates (8 kHz, 11 kHz, and 16 kHz). The feature vectors consist of 13 static cepstral coefficients and a log-energy coefficient.

4.3.2 Noise-robust front-end standard ES 202 050

This standard (ES 202 050 2007) for a front end ensures compatibility between the terminal and the remote recognizer. The first ETSI standard DSR front end (ES 201 108 2000) was published in February 2000 and is based on the mel-cepstrum representation that has been used extensively in speech recognition systems. This second standard is for an advanced DSR front end that provides substantially improved recognition performance against background noise. Evaluation of the performance during the selection of this standard showed an average of 53% reduction in speech recognition error rates in noise compared to ES 201 108.

4.3.3 Tonal-language recognition standard ES 202 211

For some applications, it may be necessary to reconstruct the speech waveform at the back end. Examples include: interactive voice response services based on the DSR of 'sensitive' information, such as banking and brokerage transactions. DSR features may be stored for future human verification purposes or in order to satisfy legal requirements.

Human verification of utterances in a speech database are collected from a deployed DSR system. This database can then be used to retrain and tune models in order to improve system performance.

In order to enable the reconstruction of the speech waveform at the back end, additional parameters such as fundamental frequency (F0) and voicing class need to be extracted at the front end, compressed and transmitted. The availability of tonal parameters (F0 and voicing class) is also useful in enhancing the recognition accuracy of tonal languages, for example Mandarin, Cantonese, and Thai.

Standard ES 202 211 (2003) specifies a proposed standard for an extended front end that extends the mel-cepstrum front end with additional parameters, namely fundamental frequency F0 and voicing class. It also specifies the back-end speech-reconstruction algorithm using the transmitted parameters.

The specification covers the following components:

- the algorithm for extraction of additional parameters, namely fundamental frequency F0 and voicing class;
- the algorithm for pitch tracking and smoothing at the back end to minimize pitch errors;
- the algorithm for speech reconstruction at the back end to synthesize intelligible speech.

4.4 Transfer protocol

The ETSI Standard for DSR, ES 201 108 (2000), defines a signal-processing front end and a compression scheme for speech input to a speech recognition system. Some relevant characteristics of this ETSI DSR front-end codec are summarized below. The coding algorithm, a standard mel-cepstral technique common to many speech recognition systems, supports three raw sampling rates: 8 kHz, 11 kHz and 16 kHz. The mel-cepstral calculation is a frame-based scheme that produces an output vector every 10 ms.

After calculation of the mel-cepstral representation, the representation is first quantized via split-vector quantization to reduce the data rate of the encoded stream. Then the quantized vectors from two consecutive frames are put into a frame-pair (FP). A DSR front-end encoder inside the user terminal performs front-end speech processing and sends the resultant data to the speech engine in the form of FPs. Each FP contains two sets of encoded speech vectors representing 20 ms of original speech.

4.4.1 Signaling

A signaling protocol is a type of protocol used to identify signaling encapsulation. Signaling is used to identify the state of connection between telephones or VOIP terminals (IP telephone or PCs or VoWLAN units). A few examples of signaling protocols are SIP (session initiation protocol), SS7 (signaling system #7), DTMF (dual-tone multi-frequency) and ISDN (integrated services digital network).

4.4.2 RTP payload format

Real-time transport protocol (RTP) defines a standardized packet format for delivering audio and video over IP networks (Wikipedia 2011a). RTP is used extensively in communication and entertainment systems that involve streaming media, such as telephony, video teleconference applications and web-based push-to-talk features.

RTP is used in conjunction with the RTP control protocol (RTCP). While RTP carries the media streams (e.g. audio and video), RTCP is used to monitor transmission statistics and quality of service and aids synchronization of multiple streams. When both protocols are used in conjunction, RTP is originated and received on even port numbers and the associated RTCP communication uses the next higher odd port number.

RTP is one of the technical foundations of voice over IP and in this context is often used in conjunction with a signaling protocol, which assists in setting up connections across the network. RTP was developed by the Audio-Video Transport Working Group of the Internet Engineering Task Force (IETF) and was first published in 1996 as RFC 1889, superseded by RFC 3550 in 2003.

RTP is designed for end-to-end, real-time transfer of stream data. The protocol provides facilities for jitter compensation and detection of out-of-sequence arrival of data, which is common during transmissions on an IP network. RTP is regarded as the primary standard for audio/video transport in IP networks and is used with an associated profile and payload format.

Real-time streaming applications require timely delivery of information and can tolerate some packet loss to achieve this goal. For example, loss of a packet in an audio application may result in loss of a fraction of a second of audio data, which can be made unnoticeable with suitable error concealment algorithms. The transmission control protocol (TCP), although standardized for RTP use, is not normally used in RTP applications because TCP favors reliability over timeliness. Instead, the majority of the RTP implementations are built on the user datagram protocol (UDP).

Format for DSR

There is a standard that specifies an RTP payload format for encapsulating ES 201 108 (2000) front-end signal processing feature streams for DSR systems. This is RFC 3557 (2003).

An ES 201 108 DSR RTP payload datagram consists of a standard RTP header (Schultzrinne *et al.* 2003) followed by a DSR payload. The DSR payload itself is formed by concatenating a series of ES 201 108 DSR FPs. The number of FPs per payload packet should be determined by the latency and bandwidth requirements of the DSR application using this payload format. In particular, using a smaller number of FPs per payload packet in a session will result in lowered bandwidth efficiency due to the RTP/UDP/IP header overhead, while using a larger number of FPs per packet will cause longer end-to-end delay and hence increased recognition latency. Furthermore, carrying a larger number of FPs per packet will increase the possibility of catastrophic packet loss; the loss of a large number of consecutive FPs is a situation most speech recognizers have difficulty dealing with. Therefore, it is recommended that the number of FPs per DSR payload packet be minimized, subject to meeting the application's requirements on network bandwidth efficiency.

The format of the FPs is important. Pairs of the quantized 10-ms mel-cepstral frames *must* be grouped together and protected with a 4-bit CRC, forming a 92-bit FP. The length of each frame is 44 bits representing 10 ms of voice. The mel-cepstral frame formats are used when forming an FP.

4.5 Energy-aware distributed speech recognition

Since mobile devices are generally limited by computation, memory, and battery energy, so performing high-quality speech recognition on an embedded device is a challenge. Delaney *et al.* (2005) have studied the energy consumption of DSR on the HP Labs SmartBadge IV embedded system and have proposed optimizations at both the application and network layers to reduce the overall energy consumption.

The SmartBadge contains a 206 MHz StrongARM-1110 processor, StrongARM-1111 co-processor, Flash, SRAM, PCMCIA interface, and various sensor inputs such as audio, temperature, and accelerometers. It runs the Linux operating system. The SmartBadge has speech/audio-driven I/O, so ASR can provide some level of user interaction through a voice-user interface. It supports a variety of different networking hardware options including Bluetooth and 802.11b wireless interfaces.

In order to reduce the energy consumption, the following optimizations are made.

1. *Architectural optimization*. Signal processing algorithms, such as the calculation of the mel-frequency cepstrum, are generally mathematically intensive. Due to a lack of floating-point hardware, simulations showed that the StrongARM spent over 90% of its time in floating-point emulation. The pre-emphasis filter, Hamming window and FFTs were implemented using fixed-point arithmetic.

2. *Algorithmic optimization*. Simulations revealed that most of the execution time was spent in the computation of the DFT (which is implemented as an FFT). Since speech is a real-valued signal, an N-point complex FFT can be reduced to an $N = 2$-point real FFT.

3. *Dynamic voltage scaling*. Once the code is optimized for both power consumption and speed, further savings are possible by changing the processing frequency and voltage at runtime. The StrongARM processor on SmartBadge IV can be configured at runtime by a simple write to a hardware register to execute at one of 11 different frequencies. For each frequency, there is a minimum voltage the StrongARM needs in order to run correctly but with lower energy consumption.

4.6 ESR, NSR, DSR

The three architectures, ESR, NSR and DSR, each with their own pros and cons, aim to incorporate an ASR system on mobile devices. DSR is claimed to be the best solution due to its superior performance in the presence of transmission errors and noisy environments (Isaacs and Mashao 2007).

The implementation of effective mobile ASR systems (ESR) is challenged by many border conditions (Zaykovskiy 2006). In contrast to generic ASRs, a mobile recognition system has to cope with:

- limited available storage (language and acoustic models must therefore be shortened, which leads to performance degradation);

- a tiny cache of 832 KB and small and slow RAM memory of between 1 and 32 MB (many signal processing algorithms are therefore not allowed);

- low processor clock frequency (enforcing use of suboptimal algorithms);

- no hardware-based floating-point arithmetic;
- no access to the operating system for mobile phones (no low-level code optimization is possible);
- cheap microphones (often far away from the mouth, which affects the performance substantially);
- highly challenging acoustic environments (PDAs can be used everywhere: in the car, on the street, in large halls and small rooms, introducing additive and convolutional distortions of the speech signal);
- lack of real PDA recorded speech corpora;
- high energy consumption during algorithm execution;
- and so forth.

Finally, improvements that could be made to one functional block of the system are often not possible because of their effect on other parts of the system.

A server-based architecture (Zaykovskiy 2006) where both the ASR front end and back end reside at the remote server is referred to in the literature as Network Speech Recognition (NSR). Another advantage of NSR relies in the fact that it can provide access to the recognizers based on different grammars or even different languages.

A characteristic drawback of the NSR architecture is the performance degradation of the recognizer caused by using low bit-rate codecs, which becomes more severe in presence of data transmission errors and background noise.

Another important issue related to NSR design is an arrangement of the server side. In contrast to generic recognition systems, the NSR back end should be able to serve hundreds of clients simultaneously and effectively. Rose *et al.* (2003) suggest an event-driven, input–output non-blocking server framework, where the dispatcher, routing all the system's events, buffers the client's queries on the decoder proxy server, which redirects the requests to the one of the free ASR decoder processes. Such an NSR server framework might be composed of a single 1-GHz proxy server and eight 1-GHz decoder servers, each running four decoder processes, and could serve up to 128 concurrent clients.

Even though both DSR and NSR make use of the server-based back end, there are substantial differences in these two schemes, which favor DSR. First of all the speech codecs, unlike the feature-extraction algorithms, are optimized to deliver the best perceptual quality and not to provide the lowest word error rate. Secondly, ASR does not need high-quality speech, but rather some set of characteristic parameters. Thus, it requires lower data rates – 4.8 kbit/s is a common rate for the features transmission. Third, since feature extraction is performed on the client side, higher sampling rates covering the full bandwidth of the speech signal are possible. Finally, because in DSR we are not constrained to the error-mitigation algorithm of the speech codec, better error-handling methods in terms of word error rate can be developed.

Bibliography

3GPP. (2004) Recognition Performance Evaluations of Codecs for Speech Enabled Services (SES), 3GPP TR 26.943.

Bernard, A. and Alwan, A. (2002) Low-bitrate distributed speech recognition for packet-based and wireless communication. *IEEE Trans. Speech Audio Process*, 10, 570–579.

Bossert, M. (2000) *Channel Coding for Telecommunications*. John Wiley & Sons.

Delaney, B. Simunic, T. and Jayant, N. (2005) Energy aware distributed speech recognition for wireless mobile devices. *IEEE Design & Test of Computers*, 22, 39–49.

ES 201 108 (2000) Speech Processing, Transmission and Quality Aspects (STQ), distributed speech recognition; front-end feature extraction algorithm; compression algorithms. http://www.etsi.org/deliver/etsi_es/201100_201199/201108/01.01.01_60/es_201108v010101p.pdf.

ES 202 050 (2007) Speech Processing, Transmission and Quality Aspects (STQ), distributed speech recognition, advanced front-end feature extraction algorithm, compression algorithms. http://www.etsi.org/deliver/etsi_es/202000_202099/202050/01.01.05_60/es_202050v010105p.pdf.

Speech Processing, Transmission and Quality Aspects (STQ); Distributed speech recognition; Extended front-end feature extraction algorithm; Compression algorithms; Back-end speech reconstruction algorithm, 2003. http://www.etsi.org/deliver/etsi_es/202200_202299/202211/01.01.01_50/es_202211v010101m.pdf.

ES 202 212 (2005) Speech Processing, Transmission and Quality Aspects (STQ), distributed speech recognition, extended advanced front-end feature extraction algorithm, compression algorithms, back-end speech reconstruction algorithm. http://www.etsi.org/deliver/etsi_es/202200_202299/202212/01.01.01_50/es_202212v010101m.pdf.

Haavisto, P. (1998) Audiovisual signal processing for mobile communications. In *Proc. European Signal Processing Conference*, Island of Rhodes, Greece.

Isaacs, D. and Mashao, D. J. A Tutorial on Distributed Speech Recognition for Wireless Mobile Devices. In *Proceedings of Southern African Telecommunication Networks and Applications Conference*, Mauritius, 9–13 September 2007.

Kiss, I. A comparison of distributed and network speech recognition for mobile communication systems.

In *Proceedings of the International Conference on Spoken Language Processing*, Beijing, China, 2000.

Manolis, P. (2001) *Distributed Speech Recognition Issues*. Public report and presentation, Chania, Greece.

Pearce, D. (2009) Enabling new speech driven services for mobile devices: an overview of the ETSI standards activities for distributed speech recognition front-ends. *Applied Voice Input/Output Society Conference* (AVIOS 2000), San Jose, CA, May 2000.

Picone, J. W. (1993) Signal modelling technique in speech recognition. *Proc. Of the IEEE*, 81, 1215–1247.

IETF (2003). RTP Payload Format for European Telecommunications Standards Institute (ETSI) European Standard ES 201 108 Distributed Speech Recognition Encoding http://tools.ietf.org/pdf/rfc3557.pdf.

Rose, R., Arizmendi, I. and Parthasarathy, S. (2003) An efficient framework for robust mobile speech recognition services.

In *Proceedings of the International Conference on Acoustic Speech and Signal Processing*, 1, 316–319.

Sharma, S., Ellis, D., Kajarekar, S., Jain, P. and Hermansky, H. (2000) Feature extraction using non-linear transformation for robust speech recognition on the Aurora database. *IEEE International Conference on Acoustics, Speech, and Signal Processing*, 2, 1117–1120.

Sklar, B. and Harris, F. J. (2004) The ABCs of linear block codes. *IEEE Signal Process. Mag.*, 21, 14–35.

Stevens, S., Volkman, J. and Newman, E. (1937) A scale for the measurement of the psychological magnitude pitch. *Journal of the Acoustical Society of America*, 8, 185–190.

Tan, Z-H., Dalsgaard, P. and Lindberg, B. (2005) Automatic speech recognition over error-prone wireless networks. *Speech Communication*, 47, 220–242.

Waibel, A., Guetner, P., Mayfield Tomokiyo, L., Schultz, T. and Woszczyna, M. (2000) Multilinguality in speech and spoken language systems. *Proc. of IEEE*, 88, 1297–1313.

Wikipedia (2011) http://en.wikipedia.org/wiki/Real-time_Transport_Protocol.

Zaykovskiy, D. (2006) Survey of the speech recognition techniques for mobile devices. *Eleventh International Conference on Speech and Computer* (SPECOM), St. Petersburg (Russia), June 2006.

Zhang, W., He, L., Chow, Y., Yang, Z., Su, Y. (2000) The study on distributed speech recognition system. *IEEE International Conference on Acoustics, Speech, and Signal Processing*, 3, 1431–1434.

5

Context in conversation

Nitendra Rajput and Amit A. Nanavati
IBM Research, India

Recent years have seen an immense growth in the variety and volume of data being automatically generated, managed and analyzed. The emergence of sensors and the role of mobile, pervasive devices as sources of context are enabling intelligent environments. It is now becoming possible to gather various types of context – environmental, situational, personal and social – by using such sensors. We believe that the increasing availability of rich sources of context and the maturity of context aggregation and processing systems suggest that the time for creating conversational systems that can leverage context is ripe. In order to create such systems, complete user-interactive systems with dialog management that can utilize the availability of such contextual sources will have to be built.

5.1 Context modeling and aggregation

In their survey, Strang and Linnhoff-Popien (2004) review and classify numerous approches to context modeling and evaluate them. The objective of most recent research is to develop uniform context models, representation and query languages as well as reasoning algorithms that facilitate context sharing and interoperability of applications. The context modeling approaches are classified

Speech in Mobile and Pervasive Environments, First Edition.
Nitendra Rajput and Amit A. Nanavati.

by the scheme of data structures that are used to exchange contextual information in the respective system:

- *Key-value models*. The model of key-value pairs is the most simple data structure for modeling contextual information. Key-value pairs are easy to manage, but lack capabilities for sophisticated structuring that would enable efficient context-retrieval algorithms.

- *Markup-scheme models*. Common to all markup-scheme modeling approaches is a hierarchical data structure consisting of markup tags with attributes and content. In particular, the content of the markup tags is usually recursively defined by other markup tags.

- *Graphical models*. A very well known general purpose modeling instrument is the Unified Modeling Language (UML), which has a strong graphical component (UML diagrams). Due to its generic structure, UML is also appropriate to model the context.

- *Object-oriented models*. Common to object-oriented context modeling approaches is the intention to employ the main benefits of any object oriented approach – namely encapsulation and reusability – to cover parts of the problems arising from the dynamics of the context in ubiquitous environments. The details of context processing are encapsulated on an object level and hence hidden to other components. Access to contextual information is provided through specified interfaces only.

- *Logic-based models*. A logic defines the conditions on which a concluding expression or fact may be derived (a process known as reasoning or inferencing) from a set of other expressions or facts. To describe these conditions in a set of rules a formal system is applied. In a logic-based context model, the context is consequently defined as facts, expressions and rules. Usually contextual information is added to, updated in and deleted from a logic-based system in terms of facts or is inferred from the rules in the system. Common to all logic-based models is a high degree of formality.

- *Ontology-based models*. Ontologies are a promising instrument to specify concepts and interrelations. They are particularly suitable to project parts of the information describing and being used in our daily life onto a data structure utilizable by computers.

According to Henricksen *et al.* (2002), to facilitate the programming of context-aware applications, infrastructure is required to gather, manage and disseminate contextual information to applications. The development of appropriate context-modeling concepts can form the basis for such a context-management infrastructure. The modeling concepts are founded on an object-based approach in which contextual information is structured around a set of entities, each describing a physical or conceptual object such as a person or communication channel. Properties of entities, such as the name of a person or the identifier of a communication channel, are represented by attributes. An entity is linked to its attributes and other entities by uni-directional relationships known as associations. Each association originates at a single entity, called the owner of the association, and has one or more other participants. Associations can be viewed as assertions about their owning entity, and a context description can correspondingly be viewed as a set of such assertions. These context models can be specified diagrammatically in the form of a directed graph, in which entity and attribute types form the nodes, and associations are modeled as arcs connecting these nodes.

Henricksen *et al.* (2004) address the software engineering challenges associated with context-awareness, including privacy and usability concerns. The Context Modeling Language (CML), is a tool to assist designers with the task of exploring and specifying the context requirements of a context-aware application. CML provides a graphical notation for describing types of information (in terms of fact types), their classifications (sensed, static, profiled or derived), relevant quality metadata and dependencies between different types of information.

Henricksen *et al.* use the mapping of object-role modeling (Halpin 2001) to the relational model to create a relational representation of context information that is well suited to context-management tasks, such as enforcement of the constraints captured by CML, storage within a database and querying by context-aware applications. Situation abstraction is used to define high-level contexts in terms of the fact abstraction of CML. Situations can be combined, promoting reuse and enabling complex situations to be easily formed incrementally by the programmer. Situations are expressed using a form of predicate logic.

Chen and Kotz (2002) propose a graph-based abstraction for collecting, aggregating, and disseminating context information. The abstraction models context information as events, produced by sources and flowing through a directed acyclic graph of event-processing operators, and then delivered to subscribing applications. Applications describe their desired event stream as a tree of operators that aggregate low-level context information published by existing sources into the high-level context information needed by the application. The operator graph is thus the dynamic combination of all applications' subscription trees.

The challenge is to collect raw data from thousands of diverse sensors, process the data into context information, and disseminate the information to hundreds of diverse applications running on thousands of devices, while scaling to large numbers of sources, applications, and users, securing context information from unauthorized uses, and respecting individuals' privacy. The authors propose a graph abstraction for context information collection, aggregation and dissemination, and show how it meets the flexibility and scalability challenges.

Figure 5.1 sketches an evolution of alternative structures. The circles are data sources, the white squares are operators and the dark rectangles represent application-specific processing. A context-aware application attempts to adapt to its changing context by monitoring a variety of sensors. Figure 5.1(a) depicts an application receiving sensor data from three sources. The application runs on one platform, commonly a mobile or embedded host. The sensors are located in the infrastructure. Figure 5.1(b) shows that much of the processing has been moved of the application platform, and may be shared by multiple applications. The context service provider defines the semantics of the context information it provides. While it is possible that the information meets the needs of some applications, in general the applications must process the output of the context service. Alternatively, the application could push its application-specific processing into the network as a proxy, essentially an application-specific context service. Figure 5.1(c) demonstrates this approach. Figure 5.1(d) takes the approach of decomposing the context service into smaller modules that produce context information of various types and forms. Application-specific proxies may now select the most appropriate inputs to begin their processing. Figure 5.1(e) shows that when there are many applications needing context information, they may be able to share both the

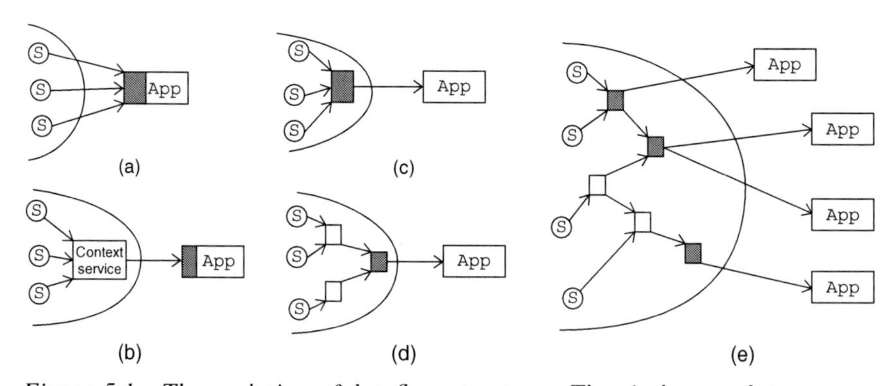

Figure 5.1 The evolution of dataflow structures. The circles are data sources, the white squares are operators, and the dark rectangles represent application-specific processing. Reproduced by permission of © 2002 IEEE.

application-specific as well as the generic processing steps. This abstraction is called an *operator* graph. The burden of converting source data into context information is on servers in the network, not on application platforms.

Context Weaver (Cohen *et al.* 2004) is a platform that simplifies writing of context-aware applications. All providers of context information registered with Context Weaver provide data to applications through a simple, uniform interface. Applications access data sources not by naming particular providers of the data, but by describing the kind of data they need. Context Weaver responds with providers of context information that may include not only devices, services, and databases external to Context Weaver, but also programmed entities that process context information from other providers. If a provider fails, Context Weaver automatically tries to rebind the application to another provider of the same kind of data. Privacy policies, specified partly by administrators and partly by individuals who are the subject of context information, are enforced by Context Weaver to protect the privacy of those individuals.

Systems that use mobile, transient, or unreliable resources typically use descriptive names (also known as intentional or data-centric names) to specify those resources. Such names identify what data is needed rather than where that data is to be found. The Context Weaver middleware for collecting and composing data from pervasive networked data providers uses a descriptive naming system. The system is based on a hierarchy of provider kinds and exploits emerging XML standards so that arbitrarily complex constraints can be specified (Cohen *et al.* 2004).

Cohen *et al.* (2002) have developed a programming model for writing context-sensitive, mobile applications, in which entities called composers accept data from one or more sources, and act as sources of higher-level data. We have defined and implemented a nonprocedural language, iQL, specifying the behavior of composers. An iQL programmer expresses requirements for data sources rather than identifying specific sources; a runtime system discovers appropriate data sources, binds to them, and rebinds when properties of data sources change. The language has powerful operators useful in composition, including operators to generate, filter and abstract streams of values.

Pervasive networked data sources, (such as web services, fixed sensors measuring traffic or weather, and mobile devices reporting position), enable context-sensitive, mobile applications (such as location monitoring, fleet management, and emergency notification). Such applications must specify how the raw data provided by networked data sources is composed into the higher-level data that it needs. Cohen *et al.* (2002) have developed a programming model and a language, named iQL, for specifying data composition rules. They have implemented the language and a runtime system that frees the application developer from many of the details that must be addressed when dealing with such data sources, including the management of widely varying protocols and formats, the discovery of appropriate data sources, and the replacement of data sources that have failed or become unreachable.

The programming model is motivated by special characteristics of pervasive networked sources, the data they provide and ways in which that data is used.

- Some data sources take the initiative in supplying data, while others do not report a value unless asked to do so.
- Pervasive data sources may fail unexpectedly, or provide inconsistent quality of service or information.
- There are often alternative ways of retrieving or deducing the same data, perhaps with different quality of service or information, from different data sources.
- Pervasive networked data sources use a wide variety of access protocols, data rates and formats.
- Raw, low-level, voluminous data, closely aligned with the characteristics of the data source, passes through a hierarchy of data-reduction transformations such as aggregation, summarization and filtering, resulting in refined, abstract, filtered data, closely aligned with the concerns of the application.
- The generation of a value by a data source can be viewed as an event. Composition of data from pervasive sources often entails composing patterns of events into higher-level compound events.
- Sensors are often deployed in arrays, providing vectors of readings conducive to data-parallel computations.

The programming model is based on entities called *composers*. A composer has a current value computed from input values, in a manner determined by a *composer specification*. Composers execute in a runtime system. A composer's input values come from *data sources*. Some data sources are pervasive networked sources such as web services and sensors, some are other composers. Data sources are *advertized* to the runtime system. A composer specification includes requirements on data sources. The runtime system discovers advertized data sources satisfying these requirements, *binds* them, and executes protocols that deliver their data to the composer. As quality-of-service and quality-of-information properties of the data source change (e.g. freshness of data, confidence in data or precision of measurement), these changes are advertized to the runtime system, which may *rebind* to different data sources that better meet the requirements.

5.1.1 An example of composer specification

Suppose workers in a one-story building wear active badges. Each badge is a data source that returns the same value – the wearer's employee number – whenever it receives a request for its current value. In addition, each badge advertizes both its employee ID number and its current location (as x and y coordinates measured in feet), and readvertizes this data every time its measured location changes. The structure of the advertisements is defined by an XML schema uniquely identified by the URL http://acmebadges.com/badgeAd. This schema defines a badge advertisement to include components named `empId` and `coordinates`. We need a list containing the employee numbers of all badge wearers within a specified distance from a specified person P, updated whenever that set of badge wearers changes (either because people move in and out of the circle around P, or because P moves, causing that circle to enclose a different set of badges). Here is the specification of a composer performing this task:

```
type Point { double x; double y; }

type EmployeeID
  schema("http://acmebadges.com/empID");

type BadgeAd
  schema("http://acmebadges.com/badgeAd");

boolean function withinDistance
  (Point p1, Point p2, double distance) {

  double dx is p2.x - p1.x;
  double dy is p2.y - p1.y;
  output dx*dx + dy*dy <= distance*distance;
}

list(EmployeeID)
  composer function AllNearbyEmployees
  (EmployeeID myID, double threshold) {

  tagged(EmployeeID) myTaggedID is
```

```
    input(BadgeAd ba:
         ba.empID=myID && ba.tagged="yes");

  BadgeAd myAd is myTaggedID.source;

  Point myPoint is myAd.coordinates;

  list(EmployeeID) nearbyIDs is
    input every
      ( BadgeAd ba:
        withinDistance
          (ba.coordinates,myPoint,threshold) );

  output nearbyIDs;
}
```

Figure 5.2 displays the expression graph corresponding to the AllNearbyEmployees specification.

Some pervasive networked data sources are *passive*, meaning that they supply current values only upon request. Others are *active*, meaning that they take the initiative in emitting a stream of values to consumers that have subscribed to them. There are also *hybrid* data sources, which take the initiative in emitting values, but also accept requests for the current (i.e. most recently emitted) value.

The iQL programming model treats all data sources as hybrid. That is, every data source, including a composer, supports two operations: requesting the current value of the data source and subscribing to notifications that the data source has generated a new value. In the case of an active data source that does not itself support the querying of a current value, the most recently generated value is taken to be the current value. A passive data source will never provide notification that it has generated a new value, but it is still possible to subscribe for such notifications.

5.2 Context-based speech applications: Conspeakuous

Conspeakuous is a context-based conversational system that explicitly manages the contextual information to be used by the spoken dialog system. It is an architecture for modeling, aggregating and using context in spoken-language conversational systems. Since Conspeakuous is aware of the environment through different sources of context, it helps in making the conversation more

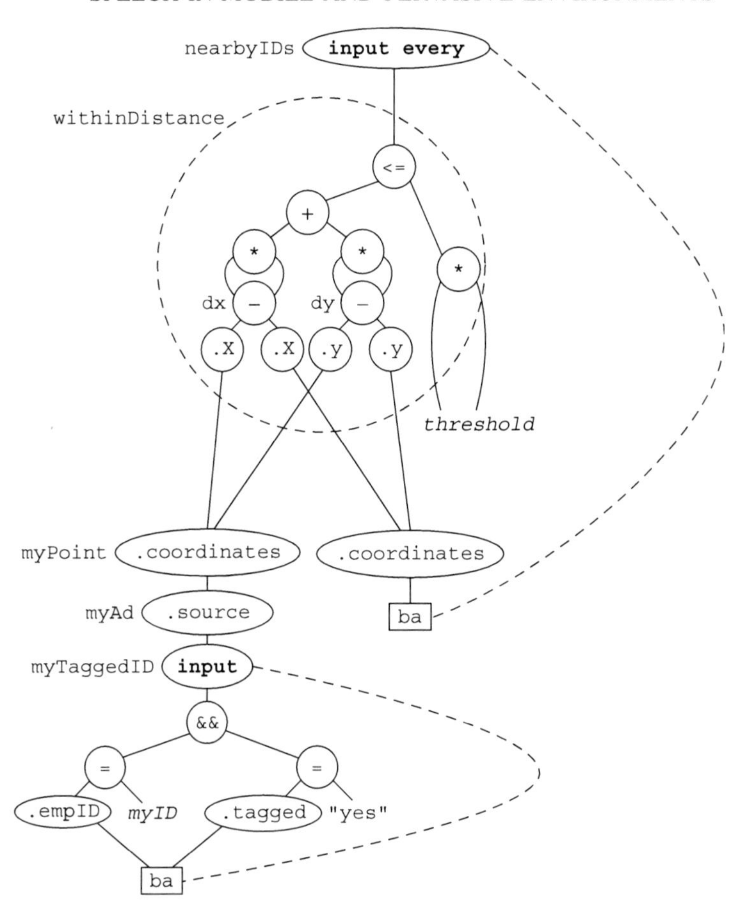

Figure 5.2 Expression graph corresponding to the AllNearbyEmployees composer specification. Reproduced by permission of © 2002 IEEE.

relevant to the user and thus reduces the cognitive load on the user. Additionally, the architecture allows for representing learning of various user/environment parameters as a source of context. An enhanced version of the architecture supports learning. In the design, learning becomes another source of context, and can therefore be composed with other sources of context to yield more refined behaviours.

Such an architecture allows building intelligent spoken dialog systems, which support the following.

- The content of a particular prompt can be changed based on the context.
- The order of interaction can be changed based on the user preferences and context.

- Additional information can be provided based on the context.
- The grammar (expected user input) of a particular utterance can be changed based on the user or the contextual information.
- Conspeakuous can itself initiate a call to the user based on a particular situation.

5.2.1 Conspeakuous architecture

Conversational systems typically do not leverage context, or do so in a limited, inflexible manner. The challenge is to design methods, systems and architectures that enable flexible alteration of dialog flow in response to changes in context.

Depending upon the dynamically changing context, the dialog task, or the very next prompt, should change. A key feature of our architecture is the separation of the context part from the conversational part, so that the context is not hard-coded and the application remains flexible to changes in context. Figure 5.3 shows the architecture of Conspeakuous (Nair *et al.* 2007). The context composer composes raw context data from multiple sources, and outputs it to a situation composer. A situation is a set or sequence of events. The situation composer defines situations based on the inputs from the context composer.

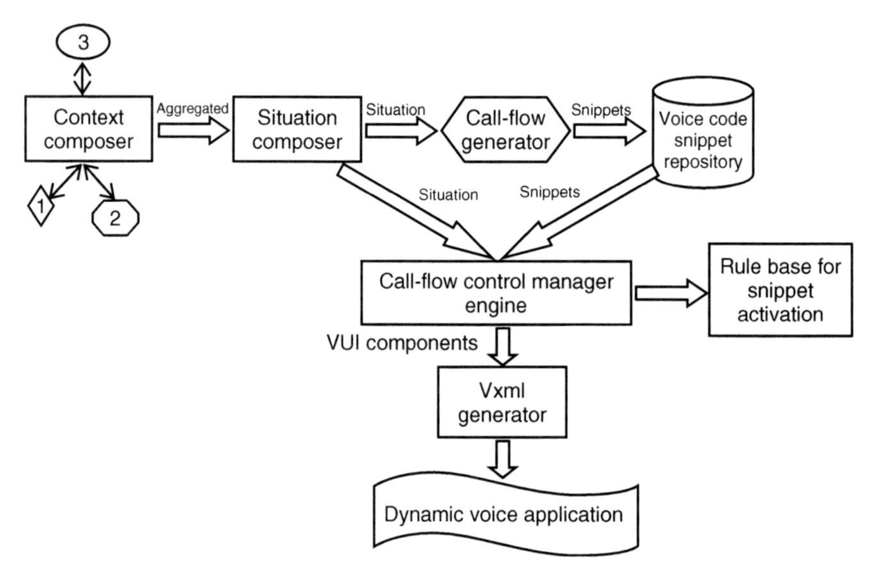

Figure 5.3 B-Conspeakuous architecture. With the kind permission of Springer Science + Business Media. © 2007 Springer.

The situations are input to the call-flow generator, which contains the logic for generating a set of dialogs (snippets) based on situations. The rule base contains the context-sensitive logic of the application flow. It details the order of snippet execution as well as the conditions under which they should be invoked. The call-flow control manager queries the rule base to select the snippets from the repository and generates the voice-user interface (VUI) components in VXML-jsp from them. We discuss two flavours of the architecture: the basic architecture B-Conspeakuous, which uses context from the external world, and its learning counterpart L-Conspeakuous, which utilizes data collected in its previous runs to modify its behaviour.

5.2.2 B-Conspeakuous

The architecture of B-Conspeakuous shown in Figure 5.3 captures the essence of contextual conversational systems. It consists of the following components.

- *Context and situation composer*. The primary function of a context composer is to collect various data from a plethora of pervasive networked devices available, and to compose it into a useful, machine-recognizable form. The situation composer composes various context sources together to define situations. For example, if one source of context is temperature and another is the speed of the wind, a composition (context composition) of the two can yield the effective wind-chill-factored temperature. A sharp *drop* in this value may indicate an event (situation composition) of an impending thunderstorm.
- *Call-flow generator*. Depending on the situation generated by the situation composer, the call-flow generator picks the appropriate voice code from the repository.
- *Call-flow control manager*. This engine is responsible for generating the presentation components to the end-user based on the interaction of the user with the system.
- *Rule-based voice snippet activation*. The rule base provides the intelligence to the call-flow control manager in terms of selecting the appropriate snippet depending on the state of the interaction.

5.2.3 Learning as a source of context

Now that a framework for adding sources of context in voice applications is in place, we can leverage this flexibility to add learning. The idea is to log all information of interest pertaining to every run of a Conspeakuous application. The logs include the context information as well as the application response. These logs can be periodically mined for 'meaningful' rules, which can be used to modify future runs of the Conspeakuous application. Although the learning module could have been a separate component in the architecture, with the context and situations as input, we prefer to model it as *another source of context*, thereby allowing the output of the learning to be further modified by composing it with other sources of context (by the context composer). This subtlety supports more refined and complex behaviours in the L-Conspeakuous application.

The L-Conspeakuous architecture is shown in Figure 5.4. It enhances B-Conspeakuous with support for closed-loop behaviour in the manner described above. It additionally consists of:

- *Rule generator*. This module mines the logs created by various runs of the application and generates appropriate association rules.
- *Rule miner*. The rule miner prunes the set of the generated association rules (further details in the next section).

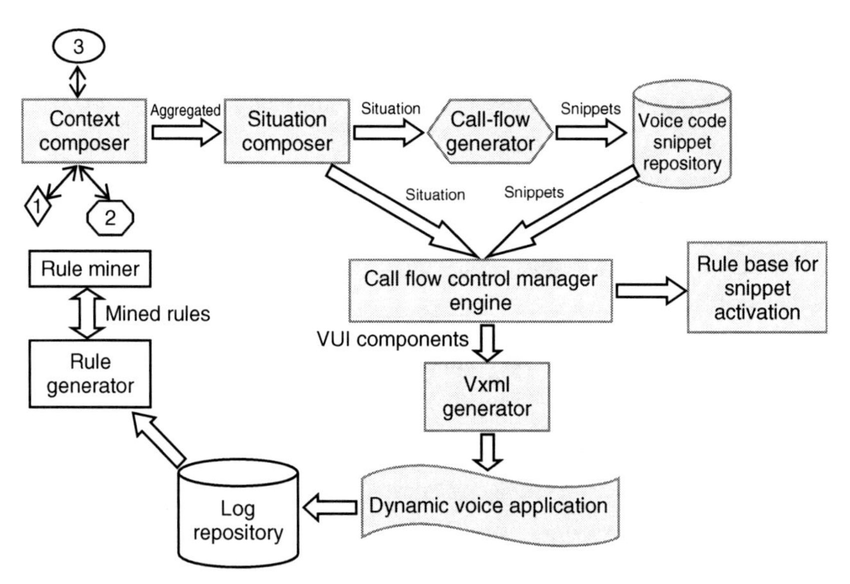

Figure 5.4 L-Conspeakuous architecture. With the kind permission of Springer Science + Business Media. © 2007 Springer.

5.2.4 Implementation

Conspeakuous has been implemented using ContextWeaver (Cohen *et al.* 2004) to capture and model the context from various sources. The development of data source providers is kept separate from voice application development. As separation between the context and the conversation is a key feature of the architecture, ConVxParser is the bridge between them in the implementation. The final application has been deployed directly on the web server and is accessed from a Genesys voice browser. The application is not only aware of its surroundings (context), but is also intelligent enough to learn from past experience. For example, it reorders some dialogs based on its learning. In the following sections we detail the implementation and working of B-Conspeakuous and L-Conspeakuous.

B-Conspeakuous implementation

With ConVxParser, the voice application developer need only add a stub to the usual voice application. ConVxParser converts this stub into real function calls, depending on whether the function is a part of the API exposed by the data provider developers or not. The information about the function call, its return type and the corresponding stub are all included in a configuration file read by ConVxParser. The configuration file (with a *.conf* extension) carries information about the API exposed by the data provider developers. For example, a typical entry in this file may look like this:

```
CON_methodname(...)
class: SampleCxSApp
object: sCxSa
```

Here, `CON_methodname(...)` is the name of the method exposed by the data provider developers. The routine is a part of the API they expose, which is supported in ContextWeaver. The other options indicate the *provider kind* that the applications need to query to get the desired data. The intermediate files, with a *.conjsp* extension, include queries to data providers in forms of stubs of pseudo code. As shown in Figure 5.5, ConVxParser parses the *.conjsp* files and using information present in the *.conf* files it generates the final *.jsp* files.

The data provider has to register a *DataSourceAdapter* and a *DataProvider-Activator* with the ContextWeaver server (Cohen *et al.* 2004). A method for interfacing with the server and acquiring the required data of a specific *provider*

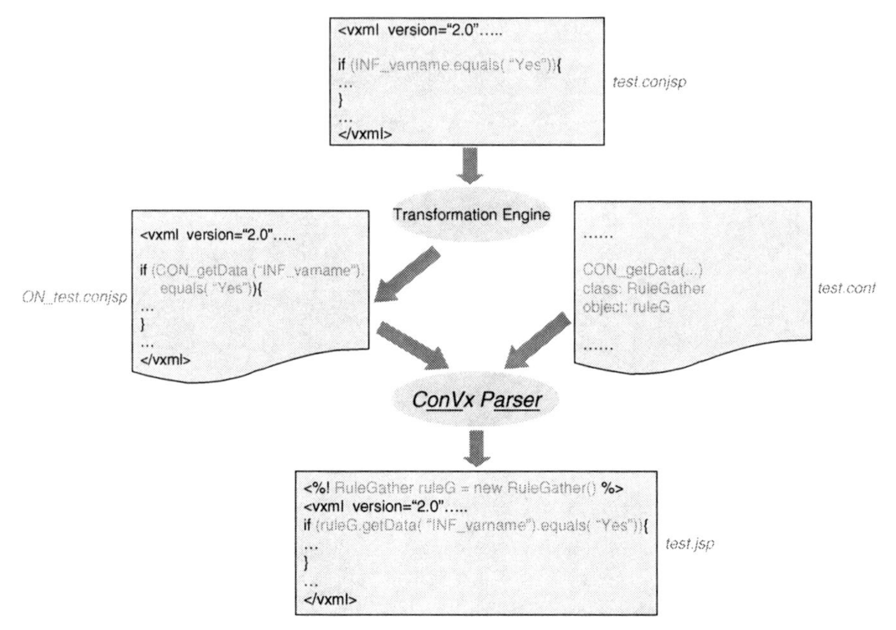

Figure 5.5 ConVxParser in B-Conspeakuous. With the kind permission of Springer Science + Business Media. © 2007 Springer.

kind, is exposed. This forms an entry of the *.conf* file with the aforementioned format. The voice application developer creates a *.conjsp* file that includes these function calls as code-stubs. The code-stubs indicate those portions of the code that we would like to be dependent on context. The input to the ConVxParser is both the *.conjsp* file and the *.conf* file. The code-stub that the application writer adds may be of the following two types. First, the method invocations are represented by `CON_methodname`, and second, the contextual variables are represented by `CONVAR_varname`.

We distinguish between *contextual* variables and *normal* variables. The meaning of a normal variable is the same as that we associate with any program variable. However, contextual variables are those that the user wants to be dependent on some source of context, i.e. one that represents real time data. There are three ways in which the contextual variables and code-stubs can be included in the intermediate voice application.

- We can assign to a *contextual variable* the output of some pseudo-method invocation for some data provider. In this case the assignment statement is removed from the final *.jsp* file but we maintain the method name–contextual variable name relation–using a HashMap so that every subsequent occurrence of that contextual variable is replaced by the appropriate method invocation. This is motivated by the fact that a contextual variable needs to be re-evaluated every time it is referenced, because it represents a real-time data.

- We can assign the value returned from a pseudo-method invocation that fetches data of some provider kind to a *normal variable*. The pseudo-method invocation to the right of such an assignment statement is converted to a real-method invocation.

- We can just have a pseudo-method invocation that is directly converted to a real-method invocation.

The data structures involved are mainly HashMaps that are used for maintaining the information about the methods described in the *.conf* file (it saves multiple parses of the file), and for maintaining the mapping between a contextual variable and the corresponding real-method invocation.

L-Conspeakuous implementation

Assuming that all information of interest has been logged, the rule generator periodically looks at the repository and generates interesting rules. Specifically, we run the *apriori* algorithm (Agrawal *et al.* 1993) to generate association rules. We modified *apriori* to support multi-valued items and ranges of values from continuous domains.

In L-Conspeakuous, we have yet another kind of variable, which we call *inferred* variables. Inferred variables are those variables whose values are determined from the rules that the rule miner generates. This requires another modification to *apriori*: only those rules that contain only inferred variables on the right-hand side are of interest to us.

The rule miner, registered as a data provider with ContextWeaver, collects all those rules (generated by the rule generator) such that:

- their left-hand sides are a superset of the current condition (as defined by the current values of the context sources);

- the inferred variable we are looking for must be in their right-hand sides.

Among the rules that are pruned out using the above stated criteria, we select the value of the inferred variable as it exists in the right-hand side of the rule with maximum support.

Figure 5.6 shows the workings in L-Conspeakuous. In addition to the code-stubs in B-Conspeakuous, the *.conjs* file has code-stubs that are used to query

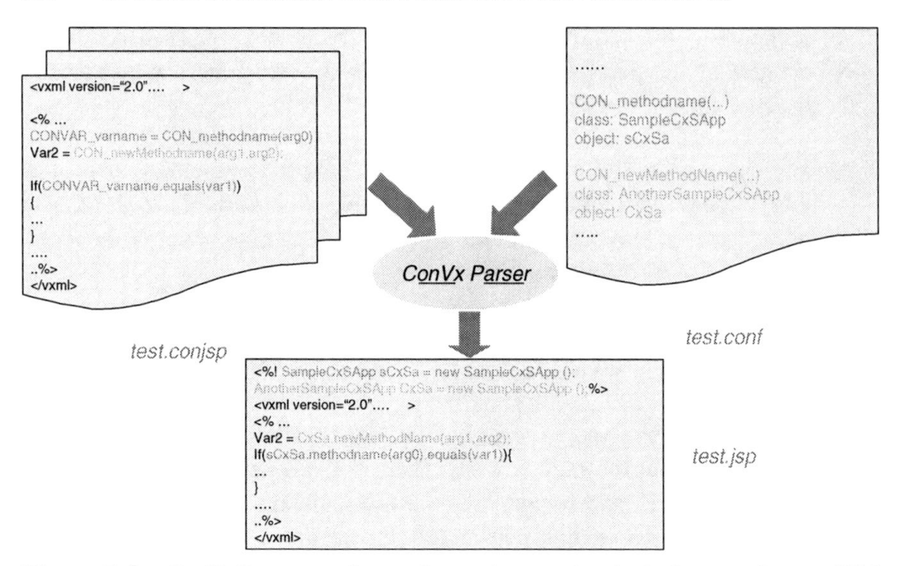

Figure 5.6 ConVxParser and transformation engine in L-Conspeakuous. With the kind permission of Springer Science + Business Media. © 2007 Springer.

values that best suit the inferred variables under current conditions. The transformation engine converts all these code-stubs into stubs that can be parsed by ConVxParser. The resulting file is a *.conjsp* file that is parsed by ConVxParser, which, along with a suitable configuration file, gets converted to the required *.jsp* file, which can then be deployed on a compatible web server.

5.2.5 A tourist portal application

We built a tourist information portal based on the Conspeakuous architecture and conducted a user study to find out the comfort level and preferences in using the B-Conspeakuous or an L-Conspeakuous systems. The application used several sources of context, which included learning as a source of context. The application suggests places to visit, depending on the current weather conditions and past user responses. The application comes alive by adding time, repeat visitor information and traffic congestion as sources of context.

The application first gets the current time from *Time DataProvider*, using which it greets the user appropriately (Good morning/evening etc.). Then, the *Revisit DataProvider* not only checks whether a caller is a re-visitor or not, but also provides the information about where it was they visited previously. If a user is a new visitor then the prompt is different from that for a re-visitor, who is asked about his previous visit. Depending on the weather conditions (from the *Weather DataProvider*) and the revisit data, the system suggests various places to visit. The list of cities is reordered based on the order of preference of previous customers. This captures the learning component of Conspeakuous. The system omits the places that the user has already visited. The chosen options are recorded in the log of the *Revisit DataProvider*. The zone from where the caller is making a call is obtained from the *Zone DataProvider*. The zone data is used along with traffic congestion information (in terms of hours to reach a place) to inform the user about the expected travel time to the chosen destination. The application was hosted on an Apache Tomcat Server, and the voice browser used is the Genesys voice portal manager.

Profile of survey subjects

Since the Conspeakuous system is intended to be used by ordinary people, we invited people such as family members, friends and colleagues to use the tourist information portal. Not all of these subjects are IT-literate, but have used some form of an interactive voice response (IVR) previously. The goal is to find whether the users prefer a system that learns user preferences or a system that is static. These were educated subjects and can converse in English. The subjects also had a fair idea of the city for which the tourist information portal had been designed. Thus the subjects had enough knowledge to verify if Conspeakuous was providing the right options based on the context and user preferences.

Survey process

We briefed the subjects for about 1 min to describe the application. The subjects were then asked to interact with the system and give their feedback on the following questions:

- Did you like the greeting that changes with the time of the day?
- Did you like the fact that the system asks you about your previous trip?
- Did you like that the system gives you an estimate of the travel duration without asking your location?
- Did you like that the system gives you a recommendation based on the current weather conditions?
- Did you like that the interaction changes based upon different situations?
- Does this system sound more intelligent that other IVRs that you have interacted with before?
- Rate the usability of this system.

User study results

Out of the six subjects that called the tourist information portal, all were able to navigate with the portal without any problems. All subjects liked the fact that the system remembered their previous interaction and conversed in that context when they called the system for the second time. Three subjects liked that the system provided an estimate of travel time without them having to provide the location explicitly. All subjects liked the fact that the system provided the best site based on the current weather in the city. Four subjects found the system to be more intelligent than all other IVRs that they had used previously. The usability scores given by the subjects were 7, 9, 5, 9, 8 and 7, where 1 is the worst and 10 is the best.

The user studies clearly suggest that the increased intelligence of the conversational system is appreciated by subjects. Moreover, subjects were even more impressed they were told that the Conspeakuous system performs the relevant interaction based on the location, time and weather. The cognitive load on the user is tremendously reduced for the amount of information that the system can provide to the subjects.

5.3 Context-based speech applications: Responsive information architect

Responsive Information Architect (RIA) can engage users in an intelligent multimodal conversation. A user can interact with RIA using multiple input channels, such as speech and gesture. To understand a user input, the multimodal interpreter exploits various contexts (e.g. conversation history) to produce an interpretation frame that captures the meanings of the input. Based on the interpretation frame, the conversation facilitator decides how RIA should act by generating a set of conversation acts (e.g. 'Describe information to the user'). Upon receiving the conversation acts, the presentation broker sketches a presentation draft that expresses the outline of a multimedia presentation. Based on this draft, the language and visual designers work together to author a multimedia blueprint that contains the details of a fully coordinated multimedia presentation. The blueprint is then sent to the media producer to be realized. To support all the components described above, an information server supplies contextual information, including domain data (e.g. houses and cities for a real-estate application), a conversation history (e.g. detailed conversation exchanges between RIA and a user), a user model (e.g. user profiles), and an environment model (e.g. device capabilities).

For interpreting multimodal user inputs, RIA uses a context-based multimodal interpretation framework called MIND (multimodal interpreter for natural dialog). MIND uses a wide variety of contexts to interpret the rich semantics of user inputs. First, MIND exploits a fine-grained semantic model that characterizes the meanings of user inputs and the overall conversation. Second, MIND employs an integrated interpretation approach that uses a wide variety of contexts (e.g. conversation history and domain knowledge). These two features enable MIND to enhance its understanding of user inputs, including ambiguous and incomplete inputs.

MIND supports three input modalities: speech, text and gesture. It uses IBM ViaVoice to perform speech recognition, and a statistics-based natural language understanding component (Jelinek *et al.*, 1994) to process natural language sentences. For gestures, the creators have developed a simple geometry-based gesture recognition and understanding component. MIND uses three types of context: conversation context, domain context and visual context.

Conversation context provides an overall history of a conversation. In an information-seeking environment, users tend to only explicitly or implicitly specify the new or changed aspects of the information of interest, without repeating what has been mentioned earlier in the conversation. Given a partial user input, required but unspecified information needs to be inferred from the conversation context. Currently, MIND applies an operation, called covering, to draw inferences from the conversation context (Chai *et al*. 2002).

The domain context provides the domain knowledge such as semantic and meta information about the application data. The domain context is particularly useful in resolving input ambiguities.

For example, to resolve the ambiguity of whether the object of interest is a city or a house, MIND uses the domain context.

As RIA provides a rich visual environment for users to interact with, users may refer to objects on the screen by their spatial (e.g. the house at the left corner) or perceptual attributes (e.g. the red house). To handle these spatial/perceptual references, MIND exploits the visual context, which logs the detailed semantic and syntactic structures of visual objects and their relations. More specifically, the automatically generated visual encoding for each object is maintained as a part of the system conversation unit in the conversation history. During reference resolution, MIND identifies potential candidates by mapping the referring expressions with the internal visual representation.

5.4 Conclusion

Context has been used in several speech processing techniques to improve the performance of the individual components. Techniques to develop context-dependent language models have been presented in Hacioglu and Ward (2001). However the aim of these techniques is to adapt language models for a particular domain. These techniques do not adapt the language model based on different context sources. Similarly, there is significant work reported in the literature that adapts the acoustic models to different channels (Tanaka *et al.* 2003), speakers (Wang *et al.* 2003) and domains (Visweswariah *et al.* 2004). However adaptation of dialog based on context has not been studied extensively.

In Gruenstein *et al.* (2005), the authors present *context-sensitive dynamic classes*, a mechanism for integrating contextual information from spoken dialog into a class n-gram language model. They exploit the dialog system's information state to populate dynamic classes, thus percolating contextual constraints to the recognizers language model in real time. They describe a technique for training a language model that incorporates context-sensitive dynamic classes, and which considerably reduces word error rate under several conditions.

In LuperFoy *et al.* (1998), the authors present an architecture for discourse processing using three different components – dialog management, context tracking and adaptation. However the context tracker maintains the context history of the dialog context and does not use the context from different context sources.

Bibliography

Agrawal, R., Imielinski, T. and Swami, A. (1993) Mining association rules between sets of items in large databases. *Proc. of ACM SIGMOD Conf. on Mgmt. of Data*, 207–216.

Chai, J., Pan, S. and M. X. Zhou. (2002) MIND: A context-based multimodal interpretation framework in conversation systems. *IEEE Int'l. Conf. on Multimodal Interfaces*, 2002, 87–92.

Chen, G. and Kotz, D. (2002) Context aggregation and dissemination in ubiquitous computing systems. *Fourth IEEE Workshop on Mobile Computing Systems and Applications*, 2002.

Cohen, N. H., Black, J., Castro, P., Ebling, M., Leiba, B., Misra, A. and Segmuller, W. (2004) Building context-aware applications with Context Weaver. Technical report, IBM Research, RC23388 (W0410-156).

Cohen, N. H., Castro, P. and Misra, A. (2004) What the meaning of what is: descriptive naming of data providers in Context Weaver. Technical report, IBM Research, RC23245 (W0406-079).

Cohen, N. H., Lei, H., Castro, P., Davis II, J. S. and Purakayastha, A. (2002) Composing pervasive data using iQL. WMCSA, *Fourth IEEE Workshop on Mobile Computing Systems and Applications, 2002*, 94.

Dutoit, T. (1996) *An Introduction to Text-To-Speech Synthesis*. Kluwer Academic Publishers, Norwell, MA, USA.

Gruenstein, A., Wang, C. and Seneff, S. (2005) Context-sensitive statistical language modeling. *Interspeech, 2005*, 17–20.

Hacioglu, K. and Ward, W. (2001) Dialog-context dependent language modeling combining n-grams and stochastic context-free grammars. *IEEE Int'l. Conf. on Acoustics Signal and Speech Processing*, 2001.

Halpin, T. A. (2001) *Information Modeling and Relational Databases: From Conceptual Analysis to Logical Design*. Morgan Kaufman, San Francisco.

Henricksen, K. and Indulska, J. (2004) A software engineering framework for context-aware pervasive computing. *IEEE International Conference on Pervasive Computing and Communications*.

Henricksen, K., Indulska, J. and Rakotonirainy, A. (2002) Modeling context information in pervasive computing systems. *IEEE Int'l. Conf. on Pervasive Computing, 2002*, 167–180.

Jelinck, F., Lafferty, J., Magerman, D., Mercer, R., Ratnapakhi A. and Roukos, S. (1994) Human Language Technology, *Proceedings of a Workshop* held at Plainsboro, New Jerey, USA, 272–277.

Lee, K.-F., Hon, H.-W. and Reddy, R. (1990) An overview of the SPHINX speech recognition system. *IEEE Transactions on Acoustics, Speech, and Signal Processing*, 38, 35–45.

LuperFoy, S., Duff, D., Loehr, D., Harper, L., Miller, K. and Reeder, F. (1998) An architecture for dialogue management, context tracking, and pragmatic adaptation in spoken dialogue systems. *Int'l. Conf. On Computational Linguistics, 1998*, 794–801.

Nair, S. A., Nanavati, A. A. and Rajput, N. (2007) *Conspeakuous: Contextualising Conversational Systems*. HCI International. (http://www.springer.com/computer/hci/book/978-3-540-73352-2)

Seneff, S. (1992) TINA: a natural language system for spoken language applications. *Computational Linguistics*, MIT Press Cambridge, MA, USA, 18, 61–86.

Smith, R. W. (1994) *Spoken Natural Language Dialog systems: a Practical Approach*. Oxford University Press, New York, USA.

Strang, T. and Linnhoff-Popien, C. (2004) A context modeling survey. *Workshop on Advanced Context Modelling, Reasoning and Management as part of UbiComp in ACM, 2004*.

Tanaka, K., Kuroiwa, S., Tsuge, S. and Ren, F. (2003) An acoustic model adaptation using HMM-based speech synthesis. *IEEE Int'l Conf. on Natural Language Processing and Knowledge Engineering, 2003*.

Visweswariah, K., Gopinath, R. A. and Goel, V. (2004) Task adaptation of acoustic and language models based on large quantities of data. *Int'l. Conf. on Spoken Lang. Processing, 2004*.

Wang, Z., Schultz, T. and Waibel, A. (2003) Comparison of acoustic model adaptation techniques on non-native speech. *IEEE Int'l. Conf. on Acoustics Signal and Speech Processing, 2003*.

6

Software: Infrastructure, standards, technologies

Nitendra Rajput and Amit A. Nanavati
IBM Research, India

6.1 Introduction

Interoperatibility plays an important role in software, and no less so in the case of mobile speech. The speech-application lifecycle on the mobile device involves the mobile operating system, the underlying network, the communication protocol and the application development environment. Enabling speech on remote client devices has implications on all four areas. Interestingly, all four areas are nascent and hence it is important to look at the developments from a mobile speech perspective. As is expected in a fast maturing area, several possible competitive solutions have appeared in a short time. For any technology to evolve in this space, it is important to focus on the interoperatibility through standards and ensure that the technology can support promising new platforms in this area.

In this chapter, we will focus on the following four areas: (a) the mobile operating system, (b) VoIP to transmit speech over the internet, (c) standard

Speech in Mobile and Pervasive Environments, First Edition.
Nitendra Rajput and Amit A. Nanavati.

voice application development languages such as VoiceXML and (d) application development platforms for mobile devices.

While there are a lot of propietary mobile operating systems, the ones that are prevalant and common are the Symbian operating system, the Windows mobile system and the recently launched Android system. An interesting aspect is to identify the support, if any is required, available to run speech applications locally on the device. Unlike a pervasive device, such as a car-navigation system, a mobile device has the minimum required for any speech application – a microphone and a speaker. However the operating system needs to provide the required device drivers for capture and processing of audio. Additionally, the operating system should be able to process audio in real-time to provide speech recognition. We will study the existing operating systems and their support to enable speech processing on devices.

While so far most of the speech applications have been accessed by a dumb telephone instrument, the scenario has changed significantly in this century. Client devices have matured from being just a speech transmitting machine and the underlying access network has also changed significantly. VoIP provides a mechanism to to transmit speech over a data channel. The telecommunication network is increasingly being fed by VoIP. End-user devices have also started to support connectivity through VoIP. Speech application access through VoIP is therefore a necessity in this changing environment.

VoiceXML has been the standard language to author speech applications. Similar to the HTML language, VoiceXML provides the voice-specific tags that are interpreted by voice browsers during an interaction with the caller. The W3C changes to the VoiceXML language have focused on providing rich features to the language, keeping pace with the supporting technologies, such as personalized speech synthesis and natural language processing. The changes have also focused on improving the ease of programming through a component-based approach. With processing now being available on mobile devices, the voice browsers on such devices should be able to interpret the enhancements in VoiceXML. Modification to such standard languages will also incorporate the changes to enable processing on client devices.

This chapter will also describe the application development environment that can enable development of speech applications for the different devices available. If M applications need to be authored for N different devices, then the number of applications that a developer needs to author increases to $M \times N$. To reduce this number, the ability to easily adapt an application for all devices is an important problem in view of the increasing number of devices.

This chapter will describe the state of the art in these four areas, and will then highlight the 'speech-on-mobile' specific issues in each one of them.

6.2 Mobile operating systems

Unlike the personal computer world, which is dominated by a maximum of three operating systems, the mobile platform area is very open. The operating systems are usually influenced by the mobile phone manufacturers, and typically each manufacturer has its own mobile platform. Since the processing and memory availability on the device vary significantly, a single mobile operating system cannot be used to support the multitude of devices. The design of a mobile operating system is significantly influenced by the power limitations, processing available per unit of the chip area and the memory available on the device. Although the processing and memory on the device have been increasing significantly over recent years, the recent trend in mobile devices tends to use the circuit area to design chips that are not power hungry, thus compromising on the ability to process speech. We survey three operating systems, which are used in the bulk of current devices, and at the same time, which are different enough to be studied separately.

In their paper, Noble *et al.* (1997) presented Odyssey, a software platform for application-aware adaptation. This approach relies on a partnership between the operating systems and applications for mobile information access. Odyssey was implemented as a set of extensions to the NetBSD operating system.

Odyssey monitors resources such as bandwidth, CPU cycles and battery power, and interacts with each application to best exploit them. For example, when high-bandwidth connectivity is lost due to a radio shadow, Odyssey detects the change and notifies relevant applications.

One of the applications Noble *et al.* built on Odyssey was a speech recognizer based on the Janus speech recognition system (Waibel 1996). Local recognition avoids network transmission and is unavoidable if the client is disconnected (Flinn and Satyanarayanan 1999). Remote recognition incurs the delay and energy cost of network communication but can exploit the CPU, memory and energy resources available to a remote server. The system also supports a hybrid mode of operation in which the first phase of recognition is performed locally, resulting in a compact intermediate representation that is shipped to the remote server for completion of the recognition. In effect, the hybrid mode uses the first phase of recognition as a type-specific compression technique that yields a factor-of-five reduction in data volume with little computational overhead. For details on the experiments and results, see Flinn and Satyanarayanan (1999).

These days, the market is replete with a wide variety of cell phones, and operating systems that run on these phones. Most of them seem to provide APIs for text-to-speech (TTS) and automatic speech recognition (ASR) capabilities.

The challenge will always remain with the limited resources that are available on the client. We will address this challenge later in this chapter.

6.3 Voice over internet protocol

The internet affects almost all aspects of current computer science research. Telecommunication networks are not untouched either. While telecommunication networks employ a dedicated end-to-end connection between the two talking parties, the internet network provides a shared connection, thus operating at lower costs. Transmitting voice packets over the internet was always attractive for economy reasons. However there were issues with respect to real-time transmission requirements for voice communication. The voice over internet protocol allowed just that – voice transmission over the data-network in real-time. Not surprisingly, more than half of telephony traffic now flows through the internet.

The current environment for voice transmission in an IP-based data network is shown in Figure 6.1. If the end-user device has support for IP connectivity, such as a computer or an IP phone, then the device can directly transmit digitized speech packets through the IP network. When the device connects to a PBX through an analog transmission line, then the PBX performs the packetization and transmits packets over the IP network. In this section, we will focus our attention to the former type of VoIP connectivity.

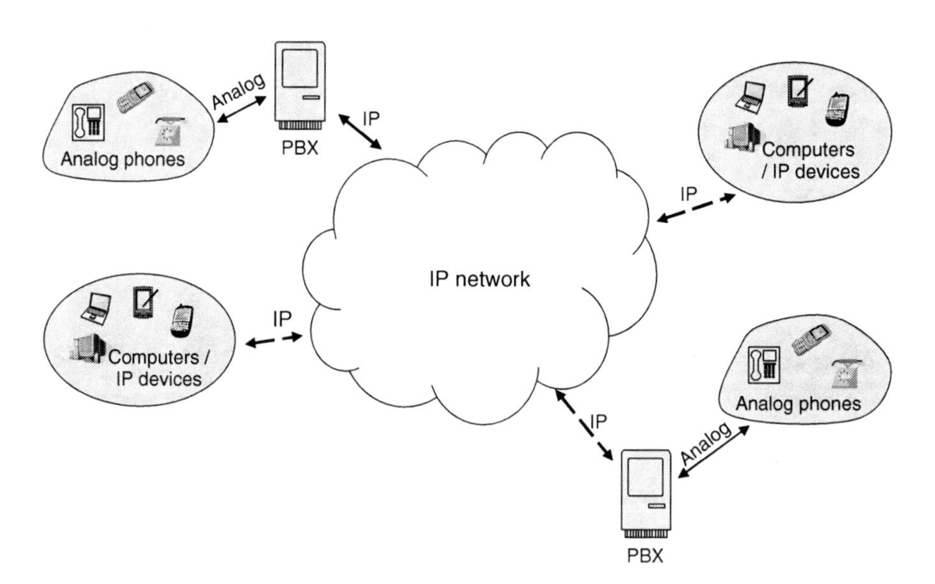

Figure 6.1 The current VoIP ecosystem.

VoIP transmits speech signals by digitizing the signals and transmitting the data-packets over the network. Unlike the telephony network transmission, the packets may traverse different paths for a particular end-to-end communication. Thus the order in which these packets arrive can be out of sequence. In a VoIP network therefore, the destination maintains a buffer of packets to generate the correct sequence before relaying them to the end-user device. Since the packets traverse through a data network, issues related to the network, such as packet loss and delay affect the quality of speech at the destination. In this section, we will study the effects of VoIP transmission on the quality of speech. We will also present some sample applications that effectively utilize the underlying data network for transmission of voice signals.

Even with the tremendous success of VoIP in telephony systems, end-user devices still operate on a analog telephony line. The last-mile connection in the mobile phone network is also over a digital circuit-switched network. As a result, a proliferation of VoIP-enabled end-user devices is yet to be seen. Such an environment requires interoperability of VoIP with the digital circuit-switched network. Later in this section, we will review the compatibility of the two networks and identify issues to be addressed in this space.

6.3.1 Implications for mobile speech

For speech to be transmitted over the IP network, it has to be compressed, encoded and then packets must be transmitted over the network. Thus speech received at the server can suffer from distortions due to compression and packet loss. Quality of service is therefore a very important factor in VoIP transmissions. The accuracy of speech recognition systems also suffers due to such transmission and compression losses in the IP networks. At one hand, it may be possible that the cost of accessing a remote voice application through the VoIP network may be significantly lower than the regular PSTN network. However, the usability of the voice application is likely to suffer due to the additional loss in the quality of speech.

Research has shown that there is a significant degradation in speech transmitted over the VoIP network. Therefore, it is important that the acoustic models in the remote voice server have been trained on VoIP acoustic data. This would account for the compression losses due to the codec. However the packet transmission loss varies with the network conditions and cannot be predicted. Therefore any remote speech application that can be accessed from a VoIP device should provide for a graceful backup interaction during any leg of the conversation. For example, a spoken dialog system that can be accessed through a VoIP device should provide for confirmations, since it is not unlikely that a particular utterance is incorrectly recognized by the system. Moreover, the dialog system should also provide a mechanism to fallback to a simpler set of dialogs that can be processed with a relatively smaller vocabulary and grammar. Such a redesign will lead to a more robust spoken dialog system that can handle packet loss over the VoIP network.

An alternative technique to avoid the compression losses due to the codec is by performing feature-extraction at the device itself, and then transmitting the features over the VoIP network. Transmission of front-end features demands a lower bandwidth than transmission of speech. Therefore the compression loss is not as significant. Significant performance improvements in speech recognition have been reported by performing frond-end processing at the client. Therefore the distributed speech recognition techniques described in the previous chapter gain even more importance when the transmission is through a VoIP device.

Even though front-end processing on the client can improve speech recognition in VoIP networks, this is not the complete solution. Not all devices may be capable of performing the processing required to generate the front-end features. Moreover, not all speech recognition systems at remote servers would be based on the same set of front-end features. So the client device needs to be configured to generate the right set of features for a specific speech recognition system. Later in this chapter, we will describe the role of standards and flexibility in performing client-side processing to counter transmission errors.

6.3.2 Sample speech applications

The most common speech application over VoIP has been the voice chat offered by several companies such as Google and Yahoo. Even though these applications do not employ a speech recognition system, they transmit voice over the IP network, and thus offer a cheaper alternative to the telephony network. The quality of speech on such applications has increased tremendously over the last two years, with the increase in the base technology of VoIP transmission standards.

Skype is yet another application that offers a VoIP connection to the PSTN network, thus allowing a soft phone to talk with a phone in the PSTN network. Skype and voice chat software provide a cost advantage and hence are able to get customers even with a lower quality of service and reliability than the regular telephony network.

6.3.3 Access channels

The VoIP channel has been penetrating the enterprise market as a solution for international calls and for easy integration of data and voice services within the enterprise. The presence of data networking infrastructure in enterprises encourages them to provide voice services using this network. At the same time, a number of companies now offer VoIP services for end users also. However, most such companies expect users to have a soft phone in their computer through which they can make voice calls using their broadband connection. As shown in Figure 6.1, access to a VoIP network is available at different levels. A call from a PSTN phone can be converted to VoIP for long-distance transmission and then converted back to the PSTN phone format. People can also use a softphone on a PC to send voice directly over the data network. Several phone companies (Cisco, DualPhone, Easy3Call) have now started to produce IP phones that can be used to transmit voice directly over the data network through various data protocols such as Session Initiation Protocol (SIP). Some manufacturers also use a traditional PSTN phone device coupled with an adapter to convert voice to data. However the penetration of these IP phones is still not very significant, when compared with the number of mobile phones.

As VoIP evolves it is expected that more and more end-devices will start transmitting VoIP data and thus the entire call setup from the caller phone to the receiver phone will be over VoIP. Until then, it is likely that access to VoIP will be available at different levels through IP phones, data adapters and PBX that can convert a PSTN call to VoIP and vice-versa.

6.4 Standards

The W3C has developed several standards to enable voice application development and deployment over telephony networks. Some of these standards are parallel to the World Wide Web world, which has the standard HTML language and HTTP protocol so that web applications can be accessed by different web browsers through a standard authoring and session language. For voice applications, the equivalent of HTML is Voice Extensible Markup Language (or VoiceXML or VXML). In addition to VoiceXML, voice applications need standards to define speech-recognition grammars and the nature of voice in synthesized prompts. Therefore, W3C has developed such standards: for representing grammars, through Speech Recognition Grammar Syntax (SRGS), and for specifying the voice output through Speech Synthesis Markup Language (SSML).

While most voice browsers support the VoiceXML and related standards provided by the W3C, other proprietary languages still exist in the market. The Speech Application Language Tags (SALT) is another standard language used to author voice applications by embedding SALT tags in other markup languages such as HTML. In this section, we will restrict our discussion to the W3C-supported standards for voice applications – VoiceXML, SRGS and SSML. We will focus on the implications for mobile speech. We will also describe an alternative – FlexVoiceXML – that can be used as a standard language for developing voice applications that can use processing on the client and the server.

6.5 Standards: VXML

A speech application can be authored using a markup language such as VoiceXML or SALT. These applications are accessed by a normal telephone. A voice browser is responsible for interpreting the application authoring language and for interacting with the telephone channel. Most commercial voice browsers (Genesys, Avaya, Nortel) support the VoiceXML language. VoiceXML is similar to HTML in structure and involves rendering tags that provide the speech input and output in an interaction (*<prompt>*, *<block>*, *<grammar>*). VoiceXML also supports phone–keypad inputs. Additionally, VoiceXML has limited tags to define the application control flow (*<submit>*, *<goto>*, *<next>*). Limited conditional statements are also provided to support this control (*<if>*, *<else>*). VoiceXML supports the concepts of subdialogs, which can form a part of the interaction within the entire application. The structure of a VoiceXML application consists of *forms* and *menus* and the various form elements (such as *<field>*, *<block>*, *<record>*, *<transfer>*) that are contained within the forms. VoiceXML also provides access to call-level information (such as the caller number or the called number) and call-level events (such as call hangup).

In addition to providing the tags to control the interaction flow, VoiceXML also uses SSXML (Speech Synthesis Markup Language)and SRGF (Speech Recognition Grammar Format). SSXML provides tags to support the *prosody* and *style* of speech output. It has tags to control the emphasis (*<emphasis>*), pause (*<break>*), and style (*<proody>*) of the speech output. The type of voice (i.e. male/female) can be specified by the *<voice>* tag. Through these tags, and more, SSML can be used to represent the synthetic voice controls in a dialog system. It can also be used provide control of the synthesized voice when a user interacts with it in a dialog system. With a *<mark>* (and a *name* attribute), the specific time when a user responds while listening to a speech output can be extracted. This enables spontaneous interactions in a dialog system.

The SRGF specifies the expected speech input from the user. Since the accuracy of a speech recognition system is not high, a dialog system uses constrained speech recognition, which restricts the recognition to a few words or phrases or sentences. The specification of this expected or desired input from a user is captured by specifying a constrained grammar in SRGF. SRGF can be specified in an XML format. It has tags to provide multiple options in which a user can provide the same input (*<one-of>*). It also has an *<item>* tag through which multiple *different* inputs are made possible. A combination of the *<item>* and *<one-of>* tags can be used to express a large number of phrases or sentences through a compact grammar.

The speech processing is done entirely at the voice browser and the voice server. The voice server has the speech technology that converts speech into text and vice-versa. The voice browser calls the voice server with the text to be synthesized, as specified in VoiceXML. To interpret a user response, the voice browser calls the voice server with a correct grammar as specified in the VoiceXML document. The grammar contains a list of all possible user responses (both valid and invalid) to a question (prompt).

6.6 Standards: VoiceFleXML

Traditionally, voice-based applications are accessed using dumb telephone devices through voice browsers that reside on the server. With the increase in the processing capabilities of pervasive devices, client-side speech processing is emerging as a viable alternative. Client-side speech processing leads to a reduction in server bottlenecks, roundtrip costs and transmission errors, as well as bandwidth and connectivity requirements. The goal therefore is to leverage a client's capabilities to the fullest. Given the variety of pervasive clients, we need a flexible way of distributing speech processing between the client and the server based on the client's capabilities and runtime considerations.

We achieved this flexibility by creating *VoiceFleXML*, an extension of VoiceXML that includes tags for specifying where the processing should be done. Furthermore, for client adaptability, *VoiceFleXML* also contains tags for merging and splitting dialogs. We also introduced a synchronization protocol *SynchP* designed for coordinating the execution between the client and the server. We also implemented a prototype voice browser that supports *VoiceFleXML* and *SynchP*. Our approach is well-suited to current voice-application deployment architectures, since it requires only minor modifications to the voice browser.

With the continuing increase in the processing capabilities of handheld devices such as PDAs and mobile phones, more and richer applications have started to reside on them. With 'smart clients', there is an opportunity to offload some of the processing to the client side. Leveraging the client side helps reduce server bottlenecks, server round-trips, as well as bandwidth and connectivity requirements. This notion of utilizing the client is well known and is practised in traditional distributed client–server applications; in the web context, JavaScript is a notable example. Before JavaScript, a trivial oversight by a user in filling a form resulted in an additional server trip.

Most speech applications are conversational agents that recognize a user utterance and convert it into text (speech to text). The text is then processed and the response communicated to the user (text to speech). Since speech interfaces have traditionally worked with dumb clients (telephones) all the speech processing is done on the server. To utilize the processing capability of the client device, some systems perform part of the speech processing on the device and the rest on the server. This setup, known as distributed speech recognition (3GPP; Ramabhadran *et al.* 2004), does initial speech processing of the user utterance on the client device and the processed signals are then passed to the voice browser. Such a setting assumes fixed client capabilities and an *a-priori* division of processing between the client and the server. This approach is quite restrictive – it does not adapt to a particular client's capabilities. Over and above the advantages of client-side processing mentioned above, client-side speech processing provides an increased benefit because transmitting speech is more error-prone (due to the low bit-rates of speech coders) and expensive (because there are more bits) (Comerford *et al.* 2001; Yang and Chen 2003). Given these facts and the number and variety of client devices, *flexible* and *adaptive* client-side processing becomes even more attractive.

The goal is to design mechanisms that enable flexible sharing of speech processing between the client and the server. Depending upon the client's processing capability and runtime conditions, the client should be able to share the processing with the server. It is important that such flexibility in specifying the sharing be achieved in a *system-* and *application-independent* manner. An obvious thought is that creating an appropriate 'scripting' language that can be executed on the client should solve this problem.

However, given the architecture of current voice browsers, the assumption that dumb clients (telephones) must be supported, and that real-time responses are required of the system, it turns out that several developments were necessary: changes in the the voice browser features and the voice browser architecture, the creation of a markup language *VoiceFleXML*, and the design of a synchronization protocol, *SynchP*, for synchronizing the execution between the client and the server. Our challenge was to design these features such that we made minimal changes to the current voice browser deployment architecture, added minimal tags to VoiceXML to enable the requisite features in *VoiceFleXML*, and kept *SynchP* lightweight.

VoiceFleXML describes the application flow and provides a rich set of tags that can be used to define a flexible distribution of processing between the client and the server. *VoiceFleXML* also contains tags for merging and splitting dialogs for automatically adapting to clients with different processing capabilities. Since the actual execution of a dialog may depend on the runtime conditions at the client/server, a synchronization mechanism is necessary. We introduced a simple lightweight protocol *SynchP* for handling the synchronization between the client and the server. *SynchP* owes its lightness to the expressivity of *VoiceFleXML*: since *VoiceFleXML* allows the detailed specification of the order and extent of processing at the client and the server, the client and server VoiceXML stubs that are generated from the same *VoiceFleXML* do not require great deal of state information to be communicated.

From a deployment perspective, our approach requires only minor modifications to the voice browser component of current architectures. We will detail the modifications necessary to current voice browsers in order to implement these features. Next, we provide an overview of speech-based systems from a browsing perspective. We then present the system architecture based on *VoiceFleXML*. The details of this markup language are described. We also describe a synchronization protocol *SynchP* designed to coordinate the execution between the client and the server. We implement a prototype voice browser that supports *VoiceFleXML* and *SynchP*. The voice browser components are described thereafter, followed by the system implementation, which is explained through examples scenarios detailing the sharing of processing under different conditions.

6.6.1 Brief overview of speech-based systems

In this section, we describe the basics of a conventional speech-based system. As seen in Figure 6.2, the speech application is deployed on a standard HTTP server. The speech application can be authored using a markup language such as VoiceXML (W3C VXML) or SALT (SALT). These applications are accessed by a normal telephone. The speech processing is done entirely at the voice browser and the voice server. The voice browser is responsible for interpreting

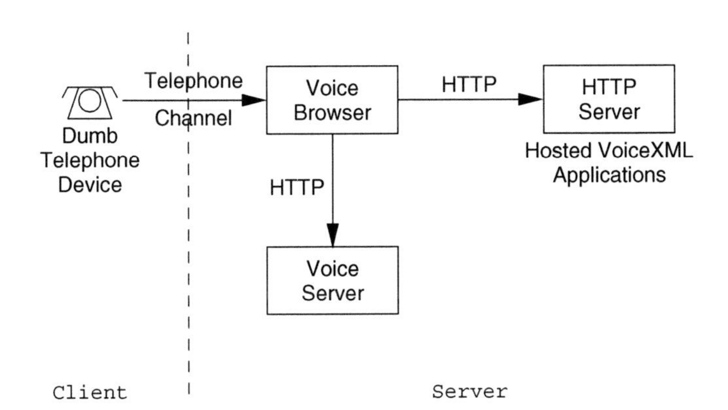

Figure 6.2 A conventional speech application deployment setting.

the VoiceXML and interacting with the telephone channel. The voice server has the speech technology that converts speech into text and vice-versa. The voice browser calls the voice server with the text to be synthesized, as specified in VoiceXML. To interpret a user response, the voice browser calls the voice server with a correct *grammar* as specified in the VoiceXML document. The grammar contains a list of all possible user responses (both valid and invalid) to a question (prompt). The voice server recognizes the user utterance with the help of the grammar and passes this to the voice browser. Thus a dumb client terminal device can be used to access the speech application.

For client devices that have the capability of executing speech applications, the voice browser and the voice server are both present locally on the client device. However, due to the processing and memory limitations on such devices, a lightweight version of a browser-cum-server is used (Comerford *et al.* 2001; Martin-Iglesias *et al.* 2005). In such cases, the entire application resides on the device and therefore only simple speech applications (with a grammar vocabulary of less than 100 words) that do not require any back-end data can be executed.

The complexity of speech applications increases with the size of the grammar vocabulary – the number of words that a user can speak while interacting with the application. For most devices with limited memory (a few tens of megabytes) and limited processing power (a few hundred megahertz), the vocabulary size is restricted to about two hundred words. Therefore the speech applications on such handheld devices are small-vocabulary applications.

6.6.2 System architecture

Supporting flexible speech processing on the client requires some changes to the traditional architecture shown in Figure 6.2. Our proposed solution is shown in Figure 6.3. On the server side, the voice server and the application server are unchanged. The traditional voice browser is replaced with a *FlexVoice* browser, which can process *VoiceFleXML*. The client communicates with the server over an HTTP connection instead of a telephone channel. The synchronization between the client and the server for processing *VoiceFleXML* is maintained using *SynchP*, which is explained in Section 6.6.4. The voice server and the application server on the client side are stripped-down versions and their capabilities and size depend on the particular client device. Depending upon their capabilities, the *FlexVoice* browser offloads computation to the server side. Note that this architecture supports dumb devices too – in this case, all the processing is done on the server side.

The voice server and the application server enable the speech-processing capability on the client. The *FlexVoice* Browser plays a central role in

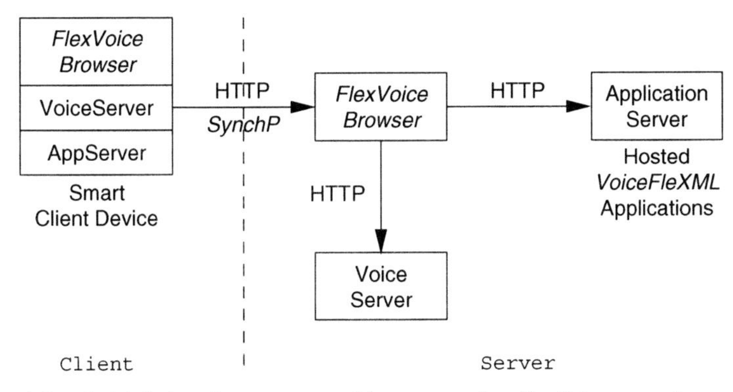

Figure 6.3 A high-level system architecture of a flexible speech processing system. The FlexVoice *browser is the voice browser enhanced with flexible speech-processing capability.*

coordinating this flexible shared processing. For this, it uses a markup language *VoiceFleXML* and the *SynchP* protocol for synchronizing the processing between the client and the server.

In this section, we describe the flexible speech processing system architecture of Figure 6.3. We derive the *VoiceFleXML* language by defining tags and attributes that are additional to those used in VoiceXML. We define the protocol, *SynchP*, which works over HTTP and maintains synchronization between the client and server execution. The *FlexVoice* browser would ideally be able to interpret the *VoiceFleXML* language and perform the processing on the client and the server. However, for the present implementation, we simulate the *FlexVoice* browser by converting *VoiceFleXML* to the client-side and server-side VoiceXMLs, and then using the standard voice browser that interprets VoiceXML. Since the client devices run on the underlying ViaVoice platform, the voice browser converts the VoiceXML to a ViaVoice-interpretable language API. This is a system-specific implementation of the *VoiceFleXML* browser, but it validates the functionality of the flexible speech-processing architecture.

This architecture provides significant flexibility in terms of where and how much processing should be done at the client as opposed to the server. This determination can also be made at runtime. If the client device is not capable of processing a complex grammar, then the architecture provides multiple options.

- The client has the flexibility of splitting the question corresponding to the complex grammar so that the underlying grammars are smaller and can be processed on the client.

- The client can ask the server to process the grammar.

Similar options are available if the speech-recognition accuracy is low for a particular utterance–grammar pair. The algorithms for splitting a grammar and the corresponding prompts of an application are described in Rajput *et al.* (2005a,b) and are shown to have a polynomial time complexity.

6.6.3 System architecture: *VoiceFleXML* interpreter

For flexibility in sharing the processing between a client and a server, we designed a markup language *VoiceFleXML*. The role of *VoiceFleXML* is two-fold: to support the specification of the location where a dialog may be processed, and to determine how much of it is processed at each location.

Since markup languages such as VoiceXML and SALT were not designed to work in a distributed environment, none of their tags and attributes support the idea of determining the location of speech processing. Also, since part of the processing is now done at the client, the memory requirement to process a particular dialog (question–answer tuple) can be reduced by splitting a dialog into various subdialogs (Rajput *et al.* 2005b). These subdialogs are less compute/memory intensive and hence can be executed on devices with limited resources. Furthermore, in order to efficiently utilize client processing capabilities, this splitting should be performed optimally (Rajput *et al.* 2005a). To enable flexibility in generating these sub-dialogs, a markup language should also provide tags that allow a device to choose the level of dialog complexity.

We describe extensions to the VoiceXML language attributes (*location* and *splitblock*) to allow the specification of the dialog processing location, namely the client or the server. We also present tags (*<split>* and *<gotomerge>*) that allow for complex speech questions to be split into smaller questions so that they can be executed on the client device. Similarly, the *<merge>* tag allows simpler questions to be combined to optimize the number of questions asked. The combinations of these tags and attributes offer several options that can be used to optimize the processing on the client and the server. We now describe the specifications of each tag and attribute of *VoiceFleXML*.

VoiceFleXML interpreter: Attributes

VoiceFleXML defines two attributes that can be used with the *<field>* and *<block>* tags of VoiceXML. These attributes can also be used with the two new tags that are described later in this section.

location. This attribute is used to specify the preferred execution location and order of the particular field or block. As seen in Table 6.1 *location* is used with the following VoiceXML tags: *<field>* and *<block>*. Additionally, this

Table 6.1 Attribute Specifications.

Attribute	Valid tags	Values
location	*<field>*	`client, server, client/server` `partial, client/partial/server`
location	*<block>*	`client, server` `client/server, server/client`
location	*<split>*	`client/split, client/server/split` `client/split/server, server/split`
location	*<merge>*	`client, client/server` `server, server/client`
splitblock	*<block>*	*num_of_chars*

attribute can be used with the *VoiceFleXML* tags *<merge>* and *<split>*, which are described later in this section. When more than one value is specified for this attribute, they identify the sequence of execution on the client and server.

Existing approaches demonstrate that the speech recognition process can be performed either on the client (Comerford *et al.* 2001; Martin-Iglesias *et al.* 2005) or on the server, or it may follow the distributed speech recognition approach (3GPP; Ramabhadran *et al.* 2004). To flexibly support the execution locations specified by these approaches, the primary values of the *location* attribute – `client`, `server` and `partial` – are used. When more than one value is specified (separated with a '/'), the first value is the preferred location. Only if the tag is unable to execute in that location is the execution performed on the next specified location. Values that are combinations of these primary values have the following implications in the execution sequence:

- `client/server`: execute on client; if the execution is unsuccessful, execute on the server.
- `client/partial/server`: execute on client; if the execution is unsuccessful, do a partial execution on client and rest on server; if this is also unsuccessful, execute on server.
- `client/split`: execute on client; if unsuccessful, split the dialog into subdialogs and then execute on the client.
- `client/split/server`: execute on client; if unsuccessful, split the dialog into subdialogs and execute on client; if this is also unsuccessful, execute the dialog on server.

- `client/server/split`: execute on client; if unsuccessful, execute on the server; if this is also unsuccessful, split the dialog into subdialogs and execute on client; if this also fails, execute the subdialogs on server.

- `server/client`: this value is valid but meaningful only to the *<merge>* and *<block>* tags; when used with *<merge>*, the merged dialog will execute on the server and if this is unsuccessful, the dialog is split and then executed on the client.

The execution can be unsuccessful for several reasons: either the client does not have enough processing capability to process the complex grammar, or the user utterance is noisy and needs a more sophisticated speech recognizer, which is not available on the client, or the client's recognition confidence is too low for the application to trust it, to mention just a few. But the flexibility provided by the *location* attribute enables use of the limited client speech-processing capabilities by having the option of transferring the processing to the server when it fails. The *VoiceFleXML* code snippet below will execute the dialog on the client side first.

```
<field name='eg1" location='client/partial/server'>
<prompt> Execute preferably on client </prompt>
<grammar src='example.grxml"/>
</field>
```

In this example, the attribute *location* has been assigned a value of `client/partial/server`. Thus the entire speech processing for this dialog will be performed at the `client`. Only for devices where this is not possible (perhaps due to memory limitations on the client) will `partial` execution be done on the client and the rest on the server. If there is a client that is not able to support even partial execution, then then entire processing is performed on the `server`. This attribute is therefore able to handle client devices of varying processing capabilities.

splitblock. The *location* attribute can be made more specific by the attribute *splitblock*. This is specifically used with the *<block>* VoiceXML tag. The *splitblock* attribute is assigned a value of the number of characters that should be processed on the client. Since this attribute is used only for the text-to-speech conversion, its use in other VoiceXML tags is not allowed. The following example shows a case where the first 18 characters of the prompt should be converted to speech at the client side and the remaining ones are converted at the server.

```
<block location='client/server' splitblock='18'>
<prompt> Execute preferably on client </prompt>
</block>
```

The *splitblock* attribute has been introduced to exploit client-side processing capabilities while asynchronous text-to-speech processing is happening on the server, an approach that is very similar in spirit to the Ajax philosophy. The client is analogous to the Ajax engine. It starts talking to the user, giving enough time to let the server to do some text-to-speech conversion and then take over the conversation with the user. The initial conversation from the client improves user response, since there is no need to wait while the text-to-speech processing takes place at the server.

VoiceFleXML interpreter: Tags

The dialogs in voice applications may have to be merged or split so that they fit in to the requirements (memory and processing capability) of the client devices. In this subsection, we illustrate the merging and splitting mechanisms and the specifications at the markup level. We describe three tags that handle the merging and splitting options for a given sequence of dialogs.

<merge>. *VoiceFleXML* uses this tag for the purpose of minimizing the number of questions that a user is asked. All the questions within this merge tag are combined into a single (complex) question. Complex speech processing is required to correctly recognize the user utterance for a merged question. Therefore the merging is performed depending on the memory availability on the client, since merging of questions increases the memory required to process the dialogs. The *<merge>* tag agglomerates the dialogs under it on the basis of memory/processing availability of the client or merges all the prompts under it to be processed on the server, depending upon the value of the *location* attribute. The final number of merged questions will be dependent on the memory available on the client device. A merge operation can be performed by the voice browser at run-time. This enables adaptation of a dialog sequence to the client resources that are available at run-time. The code snippet below shows a merge block:

```
<merge location=client/server>
<field name=month>
<prompt> Which month would you be traveling? </prompt>
<grammar src='month.grxml"/>
</field>
```

```
<field name='date">
<prompt> Which date in $month? </prompt>
<grammar src='numbers.grxml"/>
</field>
</merge>
```

Since the two *<field>* blocks in this example are inside a *<merge>* block, they can be merged depending on the client processing and memory availability. The details of the merged prompt and grammar are specified in the *<gotomerge>* tag.

<gotomerge>. All the prompts that can be required for different combinations of merging are specified in the *<gotomerge>* tag. For the example shown above, gotomerge contains the prompt that needs to be asked for a merged question. Only a *<prompt>* tag is allowed inside a *<gotomerge>* tag.

```
<gotomerge>
<prompt>pf1f2 Tell me month and day of travel.</prompt>
</gotomerge>
```

<split>. A particular question may not be executable on the client or the server due to processing limitations on the device or due to the complexity of the question. For such cases, *VoiceFleXML* provides the *<split>* tag, which can be used to represent the various options to ask the same question after splitting this into several smaller questions. The split tag uses the *location* attribute to specify the location where the split set of questions need to be executed. Typically, the client first tries to execute a complex question. If it is not successful, the client then splits this question into a few smaller questions and then tries to execute them locally. Only if this also fails does it go to a server. However several other possible mechanisms of execution can be specified using the *location* attribute with the *<split>* tag, as shown in Table 6.1. The example below shows that a question can be possibly split into three smaller questions and the corresponding prompts are also specified in the *<gotomerge>* tag.

```
<form id ='name">
<field name='location">
<prompt>Please state your city, country
   and age.</prompt>
<grammar src=location.grxml/>
<split location=client/split/server>
<merge>
```

```
<prompt>p1 Please state your city.</prompt>
<prompt>p2 Please state your country.</prompt>
<prompt>p3 Please state your age.</prompt>
<gotomerge>
<prompt>p12 Please state your city and
   country.</prompt>
<prompt>p23 Please state your country
   and age.</prompt>
</gotomerge>
</merge>
</split>
</field>
</form>
```

As seen in this example, a split set of questions can also incorporate a *<merge>* tag. This allows optimization of the split for a particular client device. By splitting and merging the dialogs in the aforementioned manner, we incorporated an ability for voice applications to adapt at run-time depending on the user's response.

VoiceFleXML has been designed to meet the requirements of a flexible client–server distributed processing architecture with minimal additions to the VoiceXML language. Only three tags are required to specify the many combinations of splitting and merging dialogs. Moreover, using two additional attributes and their possible values, *VoiceFleXML* provides the flexibility of not only choosing the speech processing location but also the preferred order of execution. Therefore with minimal extensions to VoiceXML, the architecture in Section 6.6.2 is supported.

The *<merge>* and *<split>* tags have been designed to allow *automatic* merging and splitting of prompts based on client-side constraints. The presence of tools and integrated development environments using reusable dialog components are already facilitating the creation of voice user interfaces). With the support of the automatic mechanisms for splitting and merging dialogs described in Rajput *et al.* (2005a,b), a programmer's burden is further reduced.

6.6.4 *VoiceFleXML*: Voice browser

The voice browser for *VoiceFleXML* is divided into two main components each with its specific functionalities.

1. *VoiceFleXML* interpreter

2. Synchronization engine

The *VoiceFleXML* interpreter is a minimal adaptation of existing voice browsers. The additional logic requires the interpretation of the three tags and two attributes, and it is not a significant overhead. The conversion engine is required for this specific implementation since we have used a voice server that provides a C++ API to access speech functions. The synchronization engine is also an inexpensive process as only four states are required to control the execution location and sequence. The reason for the simplicity of the synchronization protocol is that when the parser generates a client code and a server code from *VoiceFleXML*, it incorporates the complex logic of the success/failure at the server (client) into the client (server) code. The components and their interoperability are shown in Figure 6.4. At a higher level, the voice browser takes a *VoiceFleXML* file as input and generates executable files for the client and server. The synchronization protocol is also embedded in these executables. These executables are then used to run the speech application that was authored in *VoiceFleXML*. We describe each of these components in detail in this section.

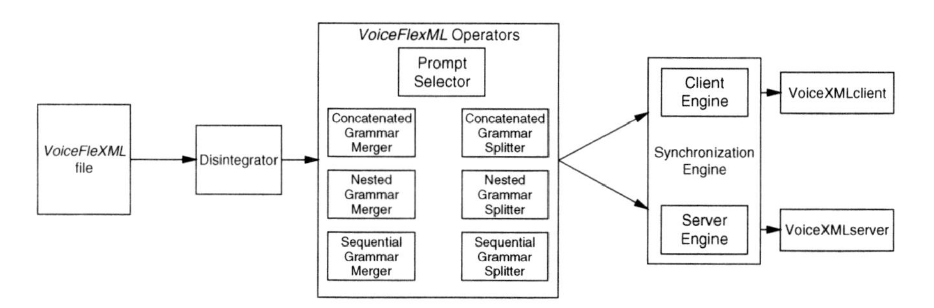

Figure 6.4 A component-based implementation of the FlexVoice *browser.*

VoiceFleXML interpreter

This layer interprets *VoiceFleXML* and generates VoiceXMLclient and VoiceXMLserver. It consists of the following components: the disintegrator and *VoiceFleXML* operators.

The disintegrator divides *VoiceFleXML* into seven parts, all of which have at most one action tag (merge/split). It routes the different parts of *VoiceFleXML* to their respective operators which will operate upon them to produce the corresponding VoiceXMLclient and VoiceXMLserver.

As for the *VoiceFleXML* operators, *VoiceFleXML* has different operators for different types of (grammar, action) pair. Considering three types of grammars (explained later in this section) and two types of actions (merge,split), we developed six operators to handle the splitting and the merging of all forms of grammars. Two operators for performing the split or merge of the prompt are also used. Each of these operators generate their specific fragments, which need to be executed on the client or the server or both. All the fragments are then integrated to form the complete VoiceXMLclient and VoiceXMLserver. We now describe each of these sub-components in detail.

Grammar splitter/merger. The grammar splitter/merger is an essential sub-component of all the main modules (excluding the blocksplitting module) of the *VoiceFleXML* interpreter layer. It handles the merging and splitting of grammars corresponding to respective dialogs in the *VoiceFleXML* file depending upon the action tag (merge,split) specified. The grammar splitter splits a grammar into smaller grammars that take less memory. The grammar merger agglomerates the given number of grammars into fewer, larger grammars, taking into account the memory constraints of the client. We categorize the grammar into three basic forms: sequential, nested and concatenated (Rajput *et al.* 2005b). The method for splitting/merging each grammar type is different. The appropriate grammar splitter/merger is invoked automatically when the corresponding grammar type appears in the *VoiceFleXML* file.

```
(a) Sequential grammar
<rule id='seatnumber">
<one-of>
<item>1</item>
<item>2</item>
</one-of>
</rule>
(b) Nested grammar
<rule id='state">
<one of>
<item><ruleref uri='#city1"></item>
<item><ruleref uri='#city2"></item>
</one of>
</rule>
(c) Concatenated grammar
<rule id='name">
<one-of>
<item>
<ruleref uri='#firstname"/>
<ruleref uri='#lastname"/>
</item>
</one-of>
</rule>
```

Prompt selector. Ideally, the new prompts should be generated by deriving information from the merged and split grammars using natural language processing. Instead, we use a *prompt selector* to identify the appropriate prompt for a given grammar operation. Therefore, the prompts for all possible merge or split dialogs have to be explicitly specified inside the *<merge>* and *<gotomerge>* tags. The prompt selector looks at the first word, namely the name of the value in the *<prompt>* tag. When grammars are merged, the name of the merged prompt is a concatenation of the respective field names. The prompt selector automatically selects the corresponding prompt based on its name.

Synchronization engine

The synchronization module is required to synchronize the processing on the client and the server. This module works in parallel with server–client engines and inserts the synchronization constraints between the executable codes. It uses *VoiceFleXML* for determining synchronization constraints. The primary states for the client and the server are *wait* and *execute*. When the client is in the *execute* state, the server is in the *wait* state and vice versa. The state diagram for the server is shown in Figure 6.5. The state diagram for the client is complementary, and therefore not shown.

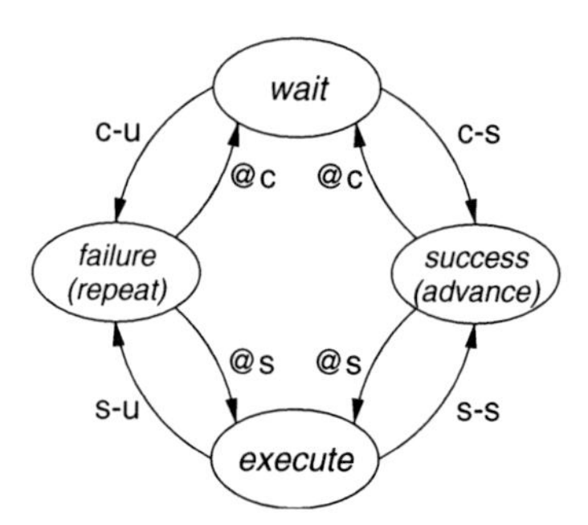

Figure 6.5 A state diagram for the server. The first letter of the labels indicates the location (c, client; s, server) of execution and the second letter indicates success or failure (s, successful; u, unsuccessful). @c(@s) means that the next question will be executed on the client (server).

Suppose the server is in the *execute* state. If the execution is successful, it sends a *successful* message to the client and moves into the *success* state and advances its pointer to the next question. If the next question is to be executed on the client, it moves into the *wait* state; otherwise it goes back into the *execute* state. If the execution is unsuccessful, it sends an *unsuccessful* message to the client and moves into the *failure* state. If the failed question is to be repeated at the client, it moves into the *wait* state, otherwise executing the question itself. With a minimal overhead of four states and the two messages, synchronization between the client and the server is achieved. The communication between the voice browser at the client and at the server is performed over an HTTP connection as is shown in Figure 6.3. When engines are to be embedded in a VoiceXML browser, the browser will have to be modified to include synchronization tags, which will be generated according to *VoiceFleXML*.

The client enters a failure state either when its speech recognition fails or if its current load does not allow processing resources. In the former case, the prompt (question) is repeated by the server and in the latter the prompt is never asked by the client. Our preliminary experiments show that the latter need not result in any perceptible delay for the user.

The possibility of user annoyance is mitigated by two factors. For one, client speech recognition is proliferating and therefore becoming more robust. Secondly, it is possible to configure the application so that it does client-side processing conservatively. This conservativeness reduces the efficiency of the system, so there is a tradeoff.

6.6.5 A prototype implementation

The *VoiceFleXML* interpreter has been implemented in Java. A DOM tree parser is used to read the *VoiceFleXML* application. This is then converted to a *VoiceXMLclient* and *VoiceXMLserver*.

The synchronization protocol is implemented in Java, and is built-in as part of the voice browser that runs on the client and the server. Thus the entire *VoiceFleXML* can be executed by such an implementation of the voice browser.

The client was an IBM ThinkPad and the server also had the same specifications. This implementation can also run on any handheld device that can execute ViaVoice applications (such as an iPaQ). We chose IBM ThinkPads for logistic simplicity.

The execution steps of the example of an airline reservation application are described in the next section. The system was tested for several *VoiceFleXML* applications. We describe the example of an air-ticket reservation system and describe the server and the client VoiceXMLs that are generated for the respective browsers. A user is prompted to specify a destination airport and the date of travel. The system then asks the user for his name and credit card details and sends the information to the back end to process the ticket. Some of these questions are processed by the server and others at the client. From a user-experience point of view, this distributed mode of processing between the client and the server remains unseen.

In Figure 6.6, the *VoiceFleXML* is shown on the left-hand box. This has six forms. Each *<form>* reflects a specific mechanism to share the processing between the client and the server. This file is passed through the *VoiceFleXML* interpreter shown in Figure 6.4 and it generates the VoiceXMLserver and VoiceXMLclient. These two VoiceXMLs share the application logic of the original *VoiceFleXML* file. The VoiceXMLclient is shown in the upper half on the

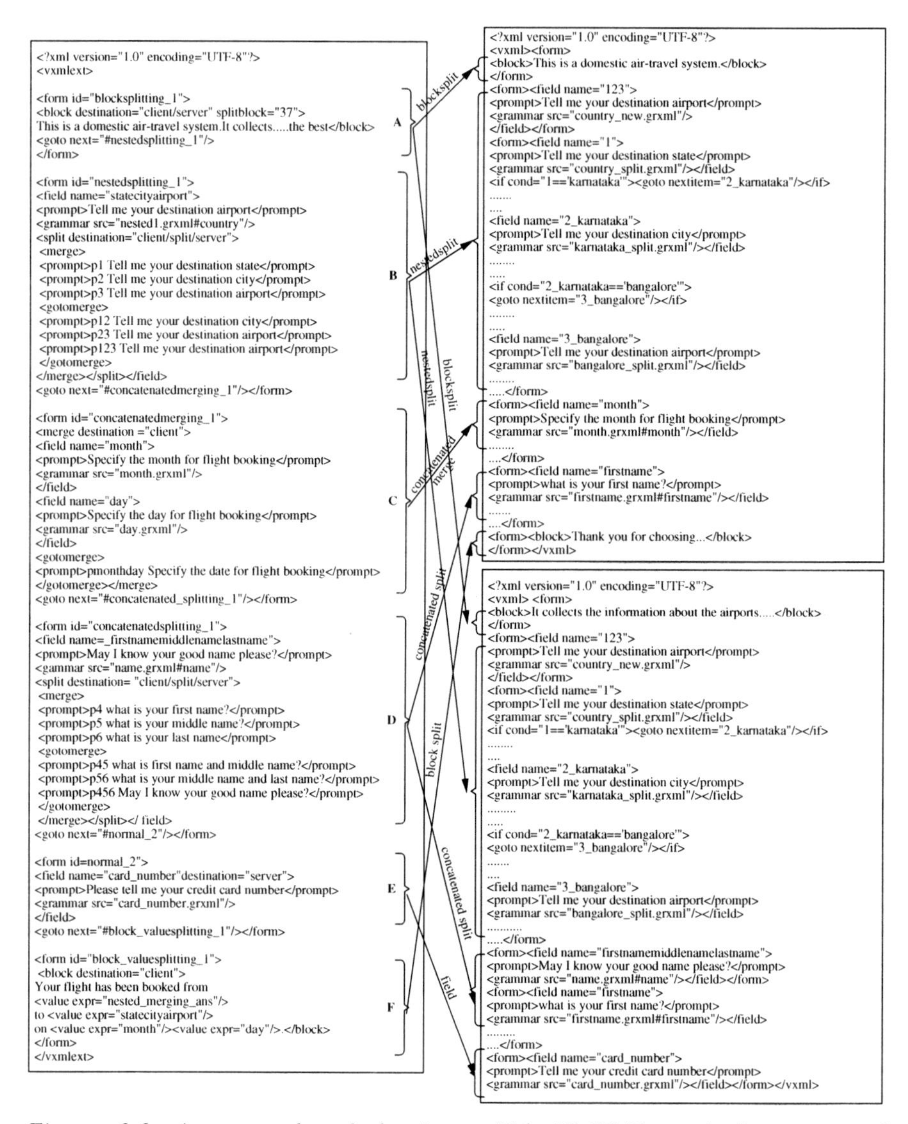

```
<?xml version="1.0" encoding="UTF-8"?>
<vxmlext>
<form id="blocksplitting_1">
<block destination="client/server" splitblock="37">
This is a domestic air-travel system.It collects.....the best</block>
<goto next="#nestedsplitting_1"/>
</form>
<form id="nestedsplitting_1">
<field name="statecityairport">
<prompt>Tell me your destination airport</prompt>
<grammar src="nested1.grxml#country"/>
<split destination="client/split/server">
<merge>
<prompt>p1 Tell me your destination state</prompt>
<prompt>p2 Tell me your destination city</prompt>
<prompt>p3 Tell me your destination airport</prompt>
<gotomerge>
<prompt>p12 Tell me your destination city</prompt>
<prompt>p23 Tell me your destination airport</prompt>
<prompt>p123 Tell me your destination airport</prompt>
</gotomerge>
</merge></split></field>
<goto next="#concatenatedmerging_1"/></form>
<form id="concatenatedmerging_1">
<merge destination ="client">
<field name="month">
<prompt>Specify the month for flight booking</prompt>
<grammar src="month.grxml"/>
</field>
<field name="day">
<prompt>Specify the day for flight booking</prompt>
<grammar src="day.grxml"/>
</field>
<gotomerge>
<prompt>pmonthday Specify the date for flight booking</prompt>
</gotomerge></merge>
<goto next="#concatenated_splitting_1"/></form>
<form id="concatenatedsplitting_1">
<field name=_firstnamemiddlenamelastname">
<prompt>May I know your good name please?</prompt>
<gammar src="name.grxml#name"/>
<split destination= "client/split/server">
<merge>
<prompt>p4 what is your first name?</prompt>
<prompt>p5 what is your middle name?</prompt>
<prompt>p6 what is your last name</prompt>
<gotomerge>
<prompt>p45 what is first name and middle name?</prompt>
<prompt>p56 what is your middle name and last name?</prompt>
<prompt>p456 May I know your good name please?</prompt>
</gotomerge>
</merge></split></ field>
<goto next="#normal_2"/></form>
<form id=normal_2">
<field name="card_number"destination="server">
<prompt>Please tell me your credit card number</prompt>
<grammar src="card_number.grxml"/>
</field>
<goto next="#block_valuesplitting_1"/></form>
<form id="block_valuesplitting_1">
<block destination="client">
Your flight has been booked from
<value expr="nested_merging_ans"/>
to <value expr="statecityairport"/>
on <value expr="month"/><value expr="day"/>.</block>
</form>
</vxmlext>
```

```
<?xml version="1.0" encoding="UTF-8"?>
<vxml><form>
<block>This is a domestic air-travel system.</block>
</form>
<form><field name="123">
<prompt>Tell me your destination airport</prompt>
<grammar src="country_new.grxml"/>
</field></form>
<form><field name="1">
<prompt>Tell me your destination state</prompt>
<grammar src="country_split.grxml"/></field>
<if cond="1=='karnataka'"><goto nextitem="2_karnataka"/></if>
.......
....
<field name="2_karnataka">
<prompt>Tell me your destination city</prompt>
<grammar src="karnataka_split.grxml"/></field>
.......
.....
<if cond="2_karnataka=='bangalore'">
<goto nextitem="3_bangalore"/></if>
.......
.....
<field name="3_bangalore">
<prompt>Tell me your destination airport</prompt>
<grammar src="bangalore_split.grxml"/></field>
.......
.....</form>
<form><field name="month">
<prompt>Specify the month for flight booking</prompt>
<grammar src="month.grxml#month"/></field>
.......
....</form>
<form><field name="firstname">
<prompt>what is your first name?</prompt>
<grammar src="firstname.grxml#firstname"/></field>
.......
....</form>
<form><block>Thank you for choosing...</block>
</form></vxml>
```

```
<?xml version="1.0" encoding="UTF-8"?>
<vxml> <form>
<block>It collects the information about the airports.....</block>
</form>
<form><field name="123">
<prompt>Tell me your destination airport</prompt>
<grammar src="country_new.grxml"/>
</field></form>
<form><field name="1">
<prompt>Tell me your destination state</prompt>
<grammar src="country_split.grxml"/></field>
<if cond="1=='karnataka'"><goto nextitem="2_karnataka"/></if>
.......
....
<field name="2_karnataka">
<prompt>Tell me your destination city</prompt>
<grammar src="karnataka_split.grxml"/></field>
........
.....
<if cond="2_karnataka=='bangalore'">
<goto nextitem="3_bangalore"/></if>
.......
....
<field name="3_bangalore">
<prompt>Tell me your destination airport</prompt>
<grammar src="bangalore_split.grxml"/></field>
..........
.....</form>
<form><field name="firstnamemiddlenamelastname">
<prompt>May I know your good name please?</prompt>
<grammar src="name.grxml#name"/></field></form>
<form><field name="firstname">
<prompt>what is your first name?</prompt>
<grammar src="firstname.grxml#firstname"/></field>
........
....</form>
<form><field name="card_number">
<prompt>Tell me your credit card number</prompt>
<grammar src="card_number.grxml"/></field></form></vxml>
```


Figure 6.6 An example of the input VoiceFleXML *and the generated VoiceXMLs for the client and server voice browsers.*

right-hand side of Figure 6.6 and the VoiceXMLserver is below it. As shown, each form of the *VoiceFleXML* has a corresponding representation at the client and/or server. The *VoiceFleXML* for airport (marked B in the figure) generates two corresponding markups called VoiceXMLserver and VoiceXMLclient. This is because the *VoiceFleXML* had the *<split>* tag with its *location* attribute as `client/split/server`. Meanwhile the *VoiceFleXML* for credit card (marked E in the figure) has its *location* attribute value as `client`. So only the corresponding VoiceXMLclient has this markup. The corresponding client and server markups generated from the *VoiceFleXML* are shown by joining them by arrows. The figure also shows the sub-components of the *VoiceFleXML* interpreter (name on the arrow) that generates the client and server versions.

Once the two versions of the VoiceXML are generated for the client and the server, the execution is performed with the use of *SynchP*. The execution is explained through a timing diagram in Figure 6.7. The execution starts with the server being in a waiting state and the client executes the welcome block (A). Once the client is done with processing its part, it sends a *successful* message to the server. The server then starts executing its part (A) and the client moves to a *wait* state. Once this is done, the client starts executing the airport form (B), but if it does not recognize the complex user utterance, it sends a *unsuccessful* message to the server and then executes the split version of that dialog (Bs). The server still is in the *waiting* state. If the client succeeds, it sends another message to the server and the execution proceeds in this manner.

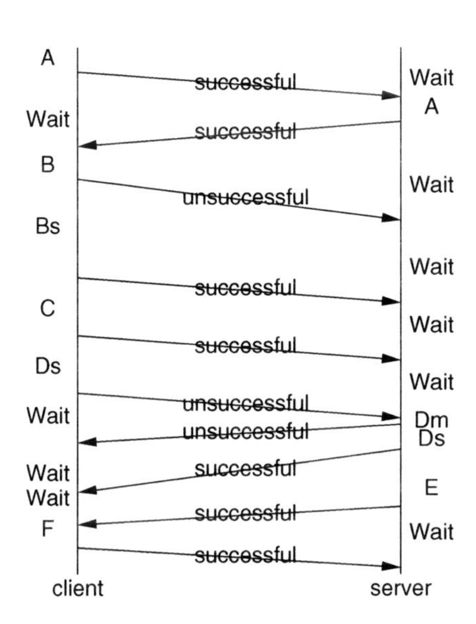

Figure 6.7 The timing diagram for the execution of the example in Figure 6.6.

When the server sends a *successful* event after the credit card form (E), the client moves to the next dialog (F) and processes the block to know about the confirmation of the ticket purchase. The example demonstrates the working of a real-world application that is deployed on the flexible speech processing architecture through the use of *VoiceFleXML* and the voice browser.

Observations

Delay. While running the proof-of-concept prototype on an IBM Thinkpad, we gave incorrect responses in order to simulate situations that result in a failure of recognition at the client. This did not result in any perceptible delay that would be annoying to a user, which is indicative. One reason we may not encounter significant delays is that the bandwidth required to transmit an 8 kHz speech signal is only 8000 × 8 bits/second, which is 64 kbps. Such bandwidth is easily available, even in the typical dialup connections.

Server load. For the example shown in Figure 6.6, two out of the six dialogs always get executed on the client, and at most five may get executed on the client. Thus the server load decreases by 33% or more.

In order to get a real-life evaluation, our next step is to implement this system on a PDA (such as an HP iPAQ) and conduct a user study to evaluate its performance.

We have described an architecture, a language and a protocol for distributing the computation of voice-based applications between the server and the client. The requirement and challenge for flexibility comes from two sources: the availability of processing capability on the client and the large variation in clients' capabilities. Using a markup-based approach, we allow the determination of the sharing to be dynamic, so that even if dialog processing fails at the client, the server can process it instead. Supporting this robustness requires a synchronization protocol. To keep the protocol lightweight, we do not keep all the state information of the client (server) on the server (client). Instead, this is built into the client and server codes that are generated after parsing the *VoiceFleXML*. *VoiceFleXML* is a small extension over VoiceXML, and yet provides enough expressiveness to support this flexibility.

For this technology to be effective, standardization of mechanisms similar to the ones described here will be necessary. As soon as various vendors support a standardized variant of *VoiceFleXML* and *SynchP*, interoperability will result. The generation of the client and server codes from a given *VoiceFleXML* document also needs to be standardized such that all the success/failure conditions are captured. To this end, our proposed architecture includes the necessary modifications to current architectures.

6.7 SAMVAAD

The proliferation of pervasive devices has stimulated the development of applications that support ubiquitous access via multiple modalities. Since the processing capabilities of pervasive devices differ vastly, device-specific application adaptation becomes essential. We address the problem of speech application adaptation by dialog call-flow reorganization for pervasive devices with different memory constraints. Given an *atomic* dialog call-flow A and device memory size m, we present optimal deterministic algorithms, RESEQUENCE and BALANCETREE, which minimize the number of questions in the reorganized output call-flow A_m. Algorithms MASQ and MATREE produce C_m, minimally distant from input call-flow C, while accommodating the memory constraint m. These two minimization criteria are capable of capturing various usability requirements important in dialog call-flow design. The following observation forms the cornerstone of all the algorithms in this chapter: two grammars g_1 and g_2 comprising $|g_1|$ and $|g_2|$ elements, respectively, can be merged into a single grammar $g = g_1 \times g_2$ having $|g_1| \cdot |g_2|$ elements for the sequential case, and $g = g_1 + g_2$ having $|g_1| + |g_2|$ elements for the tree case.

Device-specific considerations lead us to introduce the concept of an $\langle m, q\rangle$-*characterization* of a call-flow, defined as the set of pairs $\{(m_i, q_i)|i \in N\}$, where q_i is the minimum number of questions required for memory size m_i. Each call-flow has a unique, *device-independent* signature in its $\langle m, q\rangle$-characterization – a measure of its adaptability.

Here we describe SAMVAAD, a system that implements these algorithms on call-flows authored in VXML and containing SRGS grammars. The system was tested on an IBM voice browser using a test airline reservation system call-flow, reorganized for memories ranging from 64 MB to 210 KB. We ran an experiment with 14 users to obtain feedback on the usability of the adapted call-flows.

Samir is driving to a theater to watch a movie. He accesses the theater's IVR system from his mobile to make a reservation. He gets disconnected from the IVR system twice during the conversation. It would be nice if Samir could connect once to download the application, use it locally and transmit the final response, thus avoiding network instabilities as well as connection charges.

Users are increasingly accessing remote applications on the internet and running a plethora of local applications from their mobile devices. From the users' point of view, they would like more and more applications to be accessible via various interfaces (voice, multimodal) from their pervasive devices. Pervasive devices are different from desktop computers in two fundamental ways. One, they occur in various sizes with vastly differing capabilities, and by virtue of mobility are not always connected to the network. This combination gives rise to some very interesting challenges and possibilities.

From the application provider's point of view, an application composed on M pages to be accessed via N devices requires $M \times N$ authoring steps, and results in $M \times N$ presentation pages to be maintained. To address the application developer's nightmare, many application programming tools have been proposed (Banavar *et al.* 2004; Braun *et al.* 2004). Such tools allow the programmer to develop a generic application that is automatically adapted for various devices. So far, these techniques address problems in the visual domain only.

We are interested in device-specific adaptations of speech applications. Traditionally, speech applications run on a remote server, and several client–server interactions take place in the course of a dialog. A client–server model incurs transmission costs, and is prone to transmission errors, which could result in degraded speech recognition accuracy. The use of compression for reducing transmission costs introduces other complications (Ramaswamy *et al.* 1989). In order to circumvent such problems, speech recognition at the client offers a viable alternative.

Litman and Pan (1999) claim that the performance of a conversational system can be improved by adapting dialog behavior to individual users. Jameson (1998) discusses the cognitive aspects of conversational applications taking into consideration the user's available time and her ability to concentrate on the interaction. Dialog call-flow adaptation of a conversational system for improving the speech recognition accuracy has been addressed in Heisterkamp *et al.* (2003). Levin *et al.* (2000) use learning techniques for designing a conversational system and modeling it as a Markov decision process. However, none of these efforts take device characteristics into consideration.

We investigate the problem of dialog call-flow reorganization for pervasive devices with memory constraints. The crux of the reorganization lies in altering the memory requirement of the underlying grammar. We achieve this by continually merging *atomic* (an atomic grammar is one which cannot be split into subgrammars) grammars until the resulting grammar size can be supported by the device. We describe optimal deterministic algorithms, RESEQUENCE and BALANCETREE, which provide solutions for the two types of call-flow, *sequential* and *tree-type* respectively. Typically, call-flows are designed based on various usability criteria. When such 'ideal' reference call-flows are available, minimizing the number of *changes* to this ideal call-flow is a reasonable goal. MASQ and MATREE minimally alter a given dialog call-flow to accommodate it within a given memory constraint.

We introduce the concept of a *device-independent* characterization of dialog call-flows. This signature can be constructed by finding the set of criteria $\langle$memory, minimum number of questions$\rangle$ corresponding to each call-flow and provides a benchmark for the *adaptability* of a call-flow.

We have built a system, SAMVAAD, that takes as input a VXML dialog containing SRGS grammars and a memory size m. SAMVAAD has VXML and SRGS parsers that parse the input dialog to build a call-flow, which is input to the above algorithms. The output of the algorithms is converted back to a VXML dialog.

After discussing the background of the problem, we describe reorganization algorithms RESEQUENCE, BALANCETREE, MASQ and MATREE, and introduce ⟨m,q⟩-*characterization* of a call-flow. This will be followed by describing some experiments with the SAMVAAD system.

6.7.1 Background and problem setting

In this section we briefly discuss speech recognition systems, their grammars and memory requirements, and the need for client-side processing.

Client-side processing

In systems with intensive server-side processing, a typical approach to alleviate server bottlenecks and achieve scalability is to offload processing to the client side as much as possible. The evolution of Javascript in the context of the web is an example. Further, for pervasive devices, server connectivity comes at a cost and is not always robust. Together, these factors make a compelling case for disconnected, client-side processing of dialog call-flows.

Automatic speech recognition

An automatic speech recognition (ASR) system consists of two main components: an acoustic model and a language model. The acoustic model estimates how a given word or 'phone' is pronounced. The complexity (hence memory requirement) of an acoustic model is dependent on the training data and is fixed once the model is built. The language model provides a probabilistic estimate of the likelihood of a sequence of words. In conversational systems, a language model is represented by a speech recognition grammar. The memory requirement of a grammar depends on the number of choices that it encapsulates. Therefore, the memory requirement of a call-flow can be altered by changing its underlying grammars.

Memory optimization of ASR systems

While the available memory size on devices is increasing, so is their ability to support more and more complex conversational systems. Conversational systems range from single-word-based-recognition to phrase-recognition to complex-grammar recognition to large-vocabulary-recognition coupled with natural language understanding (NLU), in order of increasing memory requirement. Sophisticated speech-recognition tasks ('how-may-I-help-you' type of tasks, which provide a much better user experience) require more memory than is typically available on laptops ($\approx$512 Mb). Therefore it becomes necessary to adjust the complexity of the conversational system for different devices.

We introduce, formulate and analyze device-specific adaptation of dialog call-flows. Based on the algorithms presented, we have built a system to assist a speech application developer. The system outputs the call-flow adapted for different devices, but the task of designing corresponding prompts and help messages needs to be done by the developer. While the complete automation of dialog call-flow adaptation is a distant holy grail, this is a first step toward it.

6.7.2 Reorganization algorithms

We assume that the input call-flow comprises *atomic* dialogs only - in other words, dialogs that cannot be split into subdialogs, analogous to atomic grammars. Such a call-flow is called an *atomic call-flow*. An atomic call-flow has the smallest memory requirement and also the largest number of dialogs. M denotes the memory constraint.

We present two sets of algorithms. The first set consists of two algorithms RESEQUENCE and BALANCETREE that minimize the number of dialogs (questions/prompts) given an atomic dialog call-flow with respect to a given memory constraint and operate on sequential and tree-type call-flows respectively. The second set of algorithms MASQ and MATREE *minimally alter* the given reference call-flow (sequential and tree-type respectively) to accommodate the given memory constraint. A reference call-flow C^r may be the result of a myriad of considerations and serves as a guideline for reorganization. C^r is therefore a *soft* constraint; it need not be atomic. For this set of algorithms, we naturally require a notion of distance to quantify 'minimal alteration'. Apart from the memory constraint, all algorithms accommodate *reorganization* constraints.

Reorganization constraints

Reducing the number of questions improves usability and optimizes memory usage. However, grammar merges across certain subdialogs may either not be possible or not desirable due to factors such as data dependency and usability. We call these *reorganizational* constraints.

Reorganizational constraints represent *hard* constraints of two types. The first insists that a certain group of dialogs be merged, and the second that forbids a certain group of dialogs from being merged. The first set, *must-merge*, is a set of sets of dialogs that must be merged. The second set, *must-separate*, is a set of sets of dialogs that should not be merged. For example, in an airline reservation system (Figure 6.13), the grammar for the departure city cannot be merged with the grammar for the arrival city because the arrival city depends upon the departure city. From a usability standpoint, one may not want to merge the date of departure with the flight number. Another possible reason for preventing merges might be to improve recognition accuracy. Reorganizational constraints thus provide a mechanism to incorporate various practical considerations and constraints to improve the overall usability and performance of a dialog call-flow.

6.7.3 Minimizing the number of dialogs

In this section, we present RESEQUENCE and BALANCETREE to minimize the number of dialogs in a sequential and tree-type call-flow respectively while respecting the memory and reorganizational constraints.

RESEQUENCE

Two grammars g_1 and g_2 comprising $|g_1|$ and $|g_2|$ elements respectively can be merged into a single grammar $g = g_1 \times g_2$ having $|g_1| \cdot |g_2|$ elements. Figure 6.8 shows an example, where, as a result of a merge operation, the memory requirement goes up to 12 from 4.

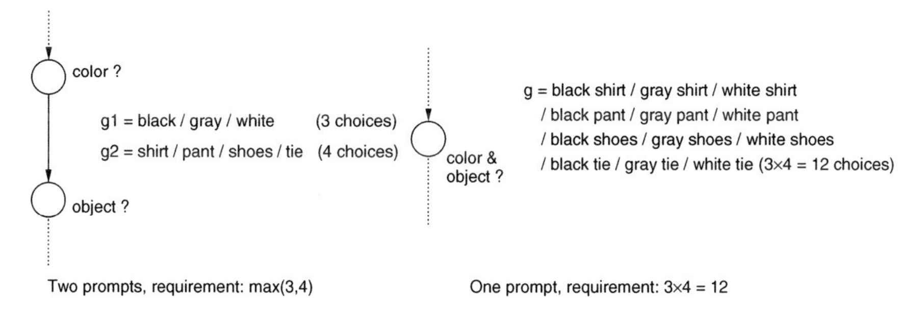

Figure 6.8 Effect of merging/splitting a sequential grammar.

A call-flow can be represented by a sequence $L = \{1, \ldots, n\}$ of *atomic* dialogs representing the order in which the dialogs are presented. The goal is to merge as many questions as possible while respecting the memory constraint. The memory requirement $m(g_i)$ for each g_i is known. We construct a graph G as follows. The vertex set $V(G)$ contains precisely the elements of L. For each vertex i in G, we add edge (i, j) if $\Pi_{k=i}^{j} m(g_k) \leq M(i < j \leq n)$, that is to say if the memory requirement of the merged grammars g_i through g_j can be accommodated within memory constraint M. As a result of this, G becomes a directed acyclic graph, possibly disconnected. Now we need to find the shortest path (or set of paths) from 1 to n, by finding the shortest path for each connected component of G. Each edge in the shortest path (set of paths) denotes the subsequence of questions being merged. The sets of dialogs in *must-merge* are merged as a preprocessing step, and dialogs merged as a result of this step are *considered atomic*. L_m denotes the output call-flow with the minimum number of dialogs. L_m may contain merged (non-atomic) dialogs. Figure 6.9 shows an example of a graph with seven nodes.

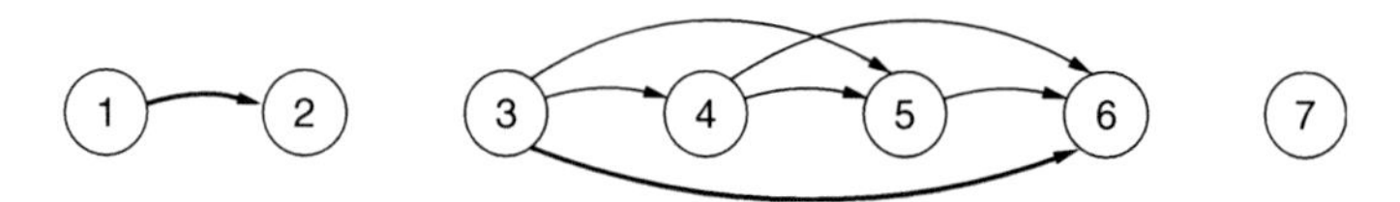

Figure 6.9 An example of a directed acyclic graph of a call-flow.

The edges of this graph represent the possible merges in the call-flow. The dotted edges identify the nodes that are not allowed to merge due to reorganization constraints. The shortest path for the graph is indicated by thick edges in Figure 6.9.

RESEQUENCE

1. input: atomic sequential call-flow L_a.
2. output: sequential call-flow L_m with the minimum number of questions.
3. Construct a graph $G(V, E)$ as follows:
   (a) Merge all *must-merge* dialogs in L_a to obtain L_a^m.
   (b) Represent all dialogs by vertices labeled $\{1, \ldots, n\}$.
   (c) For each vertex $i (1 \leq i \leq n)$
      i. for each vertex $j (i \leq j \leq n)$
      ii. if $(\Pi_{x=i}^{j} m(g_x) \leq M)$ && $\{i, j\} \notin must - separate$, add (i, j) to G.
4. Find the shortest (set of) path(s) as follows:
   (a) start = 1. $L_m = \emptyset$.
   (b) while $(start \leq n)$

i. $L_m = L_m \cup \{start\}$.

ii. select $\max_j$ such that $(start, j) \in E$.

iii. start = j+1.

5. output L_m.

RESEQUENCE is correct and has a duration of of $O(n^2)$. The graph construction phase takes a time of $O(n^2)$ to check every pair of vertices for adding edges. The shortest path phase takes a time of $O(n)$, since at each vertex the largest adjacent vertex can be chosen greedily to yield the shortest path.

BALANCETREE

Two grammars g_1 and g_2 comprising $|g_1|$ and $|g_2|$ elements respectively can be merged into a single grammar $g = g_1 + g_2$ having $|g_1| + |g_2|$ elements. Figure 6.10 shows an example. As a result of the merge operation, the memory requirement goes up from 2 to 4 ($g1$ to $g1'$).

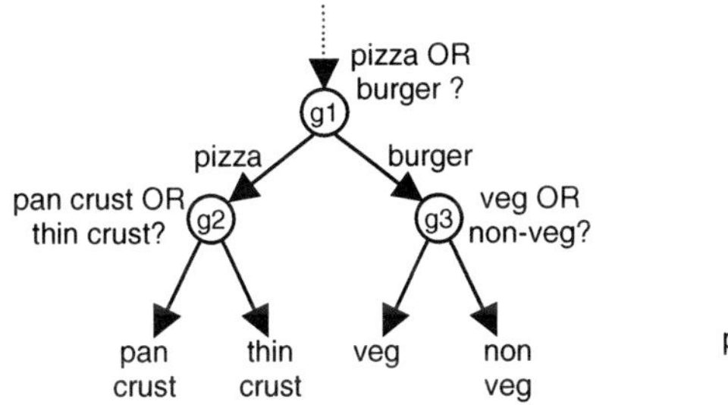

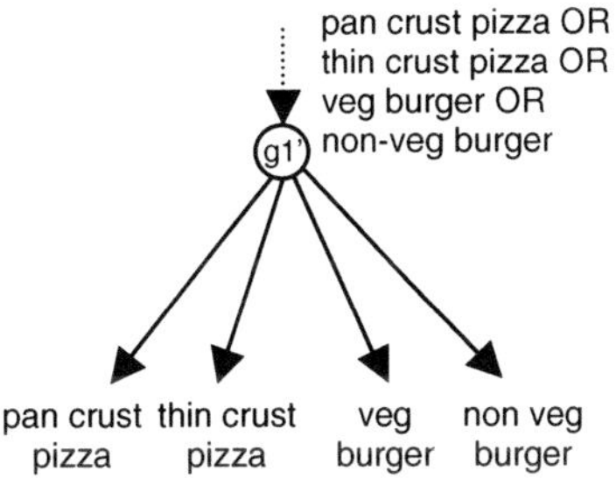

Figure 6.10 Effect of merging/splitting a tree-type grammar.

BALANCETREE

1. input: tree-type call-flow T.

2. output: tree-type call-flow T_m with the minimum number of questions.

3. initialize $T_m = T$; boolean changed = false.

4. do
   (a) Find the longest path in T_m and identify its lowest 2-subtree t_2.
   (b) if ($shorten(T_m, t_2)$) changed = true.
5. while (changed) ;
6. output T_m.
7. $shorten(T_m, t_2)$
   (a) while ($t_2 \neq$'$root$')
   (b) do
      i. if ($fold(t_2)$) return true.
      ii. else $t_2 = parent(t_2)$.
   (c) done
   (d) return false.
8. $fold(t_2)$
   (a) if (($\Delta(t_2) \geq$ degree(children(t_2))) return true.
   (b) return false.

Definition 6.7.1 The *degree* of a vertex is the number of its children.

Definition 6.7.2 A 2-subtree of a vertex v is a tree of depth 2 with v as the root.

Definition 6.7.3 A 2-subtree of a vertex v is *balanced* if all the leaves of the 2-subtree are at distance 2 from v, that is to say, no child of v is childless. A 2-subtree of a vertex v is *1-balanced* if at least one child of v is childless. A 2-subtree is either balanced or 1-balanced.

Definition 6.7.4 Let the maximum degree of any vertex in a call-flow tree be denoted by Δ. The *vacancy* of a vertex v is defined as ($\Delta - degree(v)$).

Definition 6.7.5 The *fold* operation is defined on root v of a 2-subtree and allows v to directly inherit all its grandchildren if the $\Delta \geq \Sigma_i degree(child_i(v))$. As a result of this operation, all the grandchildren of v become its own children, and the original children are removed. This operation reduces the height of the tree by 1.

The greedy application of the folding operation cannot lead to suboptimal solutions.

A greedy application of the folding operation on root v of a 2-subtree can lead to two possibilities. As a result of a *fold(v)* operation, a subsequent *fold*(*parent*(v)) is possible, or it is not. In first case, since both *fold* operations must be done optimality is preserved. In second case, it turns out that only one of *fold(v)* or *fold(parent(v))* could have been applied, either of which would lead to a height reduction of 1.

Claim 6.7.3 suggests that a bottom-up approach, on the longest paths in the tree and one 2-subtree at a time, might provide a solution. This is the essence of BALANCETREE. At each step, the longest path is found, its height reduced by 1, if a *fold* operation is possible at any vertex from the grandparent of the leaf in the longest path to the root. Note that *shorten* traverses up the tree until it is able to reduce the height by 1. After this reduction, the longest path is calculated again and the same procedure is applied. If at any time, the longest path cannot be reduced, the algorithm terminates. Since the longest path is found globally at each step, and since the height of the tree is reduced only one step at a time, we obtain a maximal height reduction. BALANCETREE is correct and runs in a time of $O(n^2)$ where n is the number of vertices in the tree. Since the *fold* operation is the dominating cost, consider the degenerate case of a tree of depth n with one vertex at each level (a path). Suppose the root has vacancy n, then each vertex folds into its parent bottom-up one at a time. This accounts for $O(n^2)$ *fold* operations. Each vertex is examined a maximum of twice for each level it visits: as a child for its degree and as a parent for its vacancy, amounting to a cost of $2n^2$.

6.7.4 Hybrid call-flows

In general, *hybrid* call-flows may contain sequential parts as well as tree-type parts. The RESEQUENCE and BALANCETREE algorithms can operate on the separate parts independently of each other. Without loss of generality, we can execute RESEQUENCE followed by BALANCETREE. As a result of RESEQUENCE, a shortened sequence may contain a vertex v with increased memory requirement and hence a reduction in *vacancy(parent(v))*. This reduction in vacancy may prevent v from folding into its parent. If RESEQUENCE would not have affected v, then BALANCETREE would have folded v into its parent. Either case leads to a height reduction of 1. This argument is similar to the one used in claim 2 above.

6.7.5 Minimally altered call-flows

Minimizing the number of questions need not be the single motivating factor for reorganization. Call-flow design entails accounting for numerous factors such as speech recognition accuracy and natural language processing. Such a well-designed call-flow can be used as a reference and altered *minimally* to meet memory constraints.

For quantifying minimality, it is necessary to define a notion of distance. We introduce a simple notion of distance based on two operations: *merge* and *split*. A single application of either of these operations on a call-flow C (whether sequential or tree-type) increases the distance of the modified version from C by 1. Let C_a denote the atomic version of C. A *split* operation on a non-atomic call-flow C can be simulated by replacing a dialog in C by its atomic components from C_a. Observe that no *split* operation was required for minimizing the number of dialogs in a call-flow because the initial call-flow was atomic. In this case, however, since the reference call-flow need not be atomic, we need to support the *split* operation.

MASQ

Given a sequential, not necessarily atomic reference call-flow L^r and a memory constraint M, MASQ constructs a call-flow L^r_m, a *minimally altered* version of L^r that satisfies the memory constraint M. MASQ is simple. If any dialog can be accommodated within M, it remains unchanged. For the others, it has to be split.

1. input: atomic sequential call-flow L_a.
2. input: reference sequential call-flow L^r.
3. output: minimally altered sequential call-flow L^r_m.
4. Construct a graph $G(V, E)$ as follows:
   (a) Represent all dialogs by vertices labeled $\{1, \ldots, n\}$
   (b) Let $L^r_m = L^r$.
   (c) for each vertex $i (1 \leq i \leq n)$
   (d) if $(m(g_i) > M)$, then $split(i, L^r_m)$.
5. output L^r_m.
6. $split(v, L^r_m)$
   (a) Find the set of atomic components $S_v = \{v_1, \ldots, v_\ell\}$ of v from L_a.
   (b) If $\exists v_i \in S_v, m(v_i) > M$, output 'IMPOSSIBLE' and exit.

(c) Otherwise,

i. $i = 1. SS_v = \emptyset$.

ii. Find the largest k such that $\Pi_i^k m(v_j) \leq M$.

iii. $i = (k + 1)$.
$SS_v = SS_v \cup \{i - k\}$.
if $(k < \ell)$, repeat previous step.

(d) Replace v by SS_v.

MASQ is correct and efficient.

The basic idea is to split only those dialogs that use memory larger than M. The *split* routine ensures the smallest number of splits. Each call to *split* involves a linear search of the corresponding atomic component set. This greedy method yields an optimal solution.

MATREE

MATREE works on tree-type call-flows. In this case, the 'splitting' of a vertex is like an 'unfolding' (similar to the *fold* operation being analogous to the *merge*). The *unfolding* operation may cause the depth of the tree to increase, but the memory requirement at the vertex decreases.

1. input: atomic tree-type call-flow T_a.
2. input: reference tree-type call-flow T^r.
3. output: minimally altered tree-type call-flow T_m^r.
4. initialize $T_m^r = T^r$;
5. do
   (a) Traverse T_m^r in preorder and if $(m(T_m^r) > M)$, then
      i. $unfold(T_m^r)$;
      ii. $matree(T_a, T_m^r)$;
6. output T_m^r.
7. $unfold(T)$
   (a) Identify the corresponding subtree T' of T in T_a.
   (b) Traverse T' bottom-up and for each vertex $v' \in T'$, $T_m' = shorten(T', v)$.
   (c) Replace T with T_m'.

MATREE produces a tree with the minimum number of alterations.

Every vertex in T_m^r can be created by merging several vertices in T_a. Each vertex in T_m^r corresponds to a unique subtree in T_a. The *unfolding* operation identifies this subtree and attempts to *shorten* it as much as possible to minimize alteration. Since each vertex corresponds to a unique subtree, the order of replacing the subtrees is inconsequential.

6.7.6 Device-independent call-flow characterization

Given a call-flow C, the above algorithms can be run with various values of memory size m_i, $1 \leq i \leq n$ and their corresponding minimum number of questions obtained. This gives us a *device-independent* characterization of C. Since these $\langle m_i, q_i \rangle$-pairs are unique for a given call-flow, they can be thought of as a *reorganizational signature* of the call-flow. We call this signature an $\langle m, q \rangle$-*characterization* of C. From a practical perspective, the $\langle m, q \rangle$-characterization of C provides a means for comparing two call-flows that essentially (semantically) perform the same task – doing an airline reservation, for example – and traces the memory requirements of each. This is important in call-flow design.

The $\langle m, q \rangle$-characterization function of C is typically a decaying function – a composition of lines with negative, decreasing slopes. Consider a sequential call-flow L of n dialogs, where each dialog i requires memory m_i, and a single question requires $\Pi_1^n m_i$. This is the largest value of the function. The smallest value is $\max(m_i)$. When all the numbers are the same, this function reduces to an exponential function on m_1. In the most general case, this function is similar to the *falling factorial* function, except that the the numbers are not necessarily consecutive, so the slope of the curve continues to decrease faster than the *falling factorial* function. In the case of tree-type call-flows, since the numbers get added rather than multiplied, the effect is less pronounced.

Figure 6.11 shows a comparison of the $\langle m, q \rangle$-characteristics of two imaginary call-flows, *Cflex* and *Cshort*. Both call-flows are semantically equivalent in that they perform the same task (for example, airline reservation) but were designed with different assumptions and considerations in mind. The choice of the call-flow could depend on a number of factors. For example, if the designer expects client devices with less than 90 bytes of memory to access the application, *Cflex* is preferable. However, if the designer is concerned that he does not want to ask more than three questions, then *Cshort* accomplishes this with less memory.

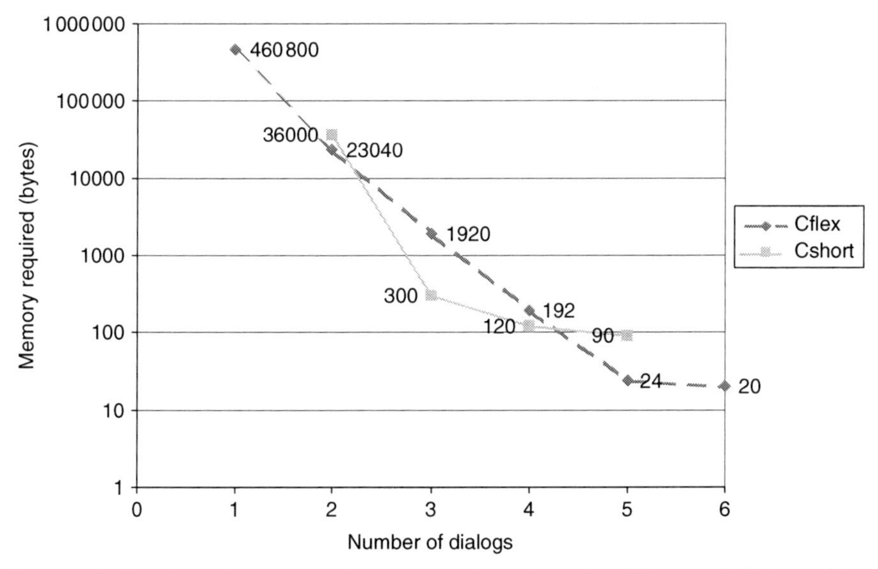

Figure 6.11 Sample $\langle m, q\rangle$-characterization plots for Cflex *and* Cshort. *Due to reorganizational constraints,* Cshort *has a minimum of two questions.* Cflex *is more flexible and can support devices with lower memories.*

6.7.7 SAMVAAD: Architecture, implementation and experiments

In this section, we detail the architecture and implementation of SAMVAAD, a dialog call-flow reorganization system. We use a sample airline reservation system call-flow to illustrate the functioning of SAMVAAD and to analyze the behavior of the algorithms for different memory constraints. We also present the outcomes of a user study we conducted to evaluate the usability of the reorganized call-flows, the output of SAMVAAD.

SAMVAAD is implemented in Java2 (v1.5.0) and the generated dialogs were deployed on an IBM WebSphere Voice Response browser that uses the IBM WebSphere Voice Server for speech recognition and speech synthesis. We tested the system on several working VXML dialogs and noted that it generated syntactically correct dialogs.

Figure 6.12 shows the architecture of SAMVAAD. A VXML dialog file specifies the dialog call-flow, its prompts and grammars. The VXML dialog file is first parsed by the VXML parser. The grammar parser parses the SRGS grammars referred to in the VXML dialog. The reorganization algorithms module processes the call-flow graph output from the VXML parser. This module

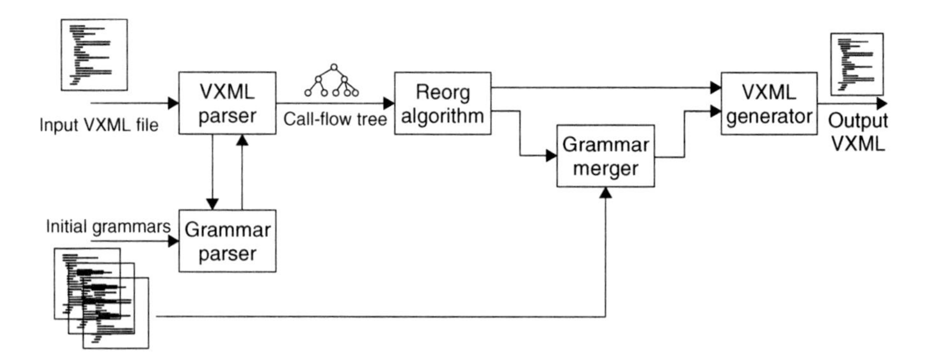

Figure 6.12 System Architecture of SAMVAAD.

outputs the reorganized call-flow graph and identifies the grammars that need to be merged. The grammar merger module merges the grammars and finally the VXML generator reconverts the call-flow graph into VXML. We now describe the five components in detail:

- *VXML parser.* This parser parses VXML (v2.0) dialogs and extracts the call-flow in a document object model (DOM) tree representation. Each node of the tree corresponds to an input element in a VXML dialog and has an associated grammar. The number of children at a node is equal to the number of choices after the input <field> block. The VXML parser outputs a sequential or tree-type call-flow. It invokes the grammar parser when required by passing a <grammar-file> handle to it.

- *Grammar parser.* This parses the grammar file associated with a node of the call-flow DOM tree. We preferred the SRGS-XML format so that we could use a JAXP implementation of the DOM parser. We parse the grammar file to count the number of choices the grammar encapsulates: if the elements of a node are present in a <one-of> block (an SRGS-XML tag element), all choices within each of these elements are *added*; otherwise, they are *multiplied*.

- *Reorganization algorithms.* This module contains the algorithms explained above.

- *Grammar merger.* Grammar merges are of two types: OR-type and the AND-type. For a OR-type merge, the final root will comprise a <one-of> block that contains references to the nodes of original grammars as its children. An AND-type grammar merge contains the rule references of the original grammars in its root node.

- *VXML generator.* The VXML generator takes the reorganized call-flow and generates the final VXML dialog that contains the merged grammars.

The VXML generator requires new prompts for the merged grammars. Our implementation of the system generates these prompts by concatenating the prompts corresponding to the original grammars. However, these can be re-authored manually to provide a better correspondence with their associated grammars. The reorganization constraints are specified in the input VXML dialog through a special reorganization tag <must-separate>, which is parsed by the VXML parser and is appropriately represented in the call-flow DOM tree structure.

To demonstrate the varying memory requirements of grammars on a device, we calculated memory required by a speech recognition system for grammars of various sizes. We used a large vocabulary English speech recognition system to decode a speech utterance. The utterance comprised a single word. The grammars used for decoding consisted of isolated words. The size of grammar therefore reflects the vocabulary of the recognition system. The decoding was performed for the same utterance, but with varying grammar sizes. Table 6.2 shows the memory required to perform decoding on a 450 MHz Quad processor AIX machine with 2 GB of RAM. These numbers may vary depending upon the particular implementation of the speech recognition system and the hardware. The memory requirement of 47916 bytes for decoding the utterance against a one-word grammar can be interpreted as the footprint that is required by the non-grammar-specific portion of the speech recognition system. The additional memory requirement with the increase of grammar size is a reflection of the increased memory requirement for decoding the same speech utterance.

Table 6.2 Memory requirement for the different grammer sizes.

Grammar size	Memory required (bytes)	Grammar size	Memory required (bytes)
1	47916	3000	53356
116	47960	4000	58712
280	48080	5000	60532
370	48052	7000	61888
960	48412	9600	62860
1440	48644	10670	63060

Figure 6.13(a) shows a sample airline reservation system call-flow. The reorganization constraints, which are of the *must-separate* type, are represented

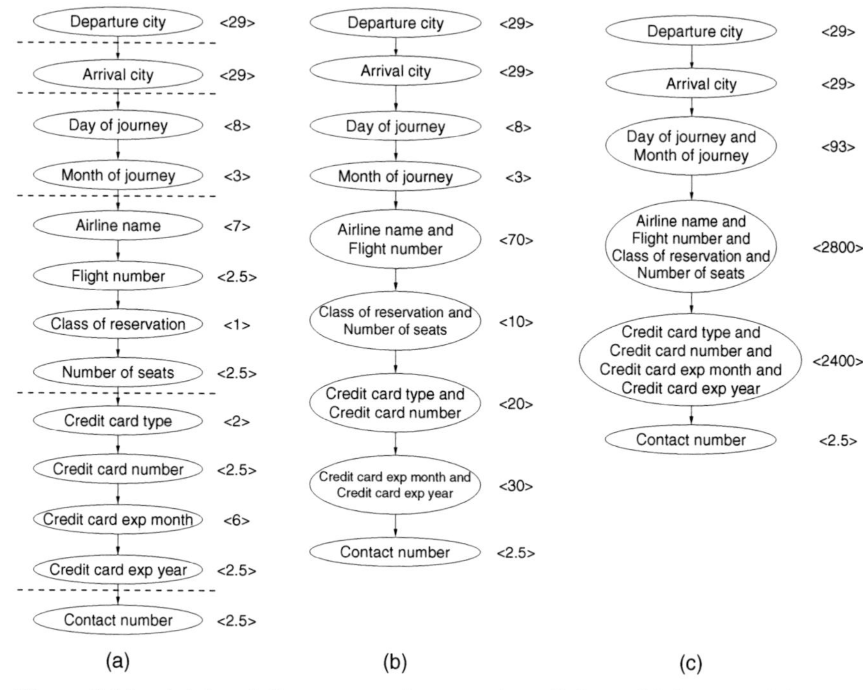

Figure 6.13 (a) An airline reservation atomic call-flow. (b) Output of the RESEQUENCE *algorithm with $m = 70$ KB. (c) Output of the* RESEQUENCE *algorithm with $m = 3400$ KB. The memory (in KB) required by ASR to process the grammar is shown in $< >$ against each dialog.*

by a dashed line in the call-flow and are extracted from the <must-separate> tag in the VXML file.

RESEQUENCE

For the input call-flow, Figure 6.13(a)–(c) show the output of RESEQUENCE for memory sizes $m = 70$ KB and $m = 2400$ KB respectively. While the first call-flow requires nine questions, the second call-flow requires six questions to gather the same information. The decrease in the number of questions is a result of merging the corresponding grammars.

The $\langle m, q \rangle$ characteristics of the above call-flow is shown in Figure 6.14. Each bar in the chart corresponds to a unique call-flow. If the call-flow has only one question, its memory requirement is too big to be accommodated by any device. At the other extreme, the call-flow that has six questions can run on all devices. The value on the y-axis refers to the memory required to execute the largest grammar in the respective call-flow. The plot has been generated by

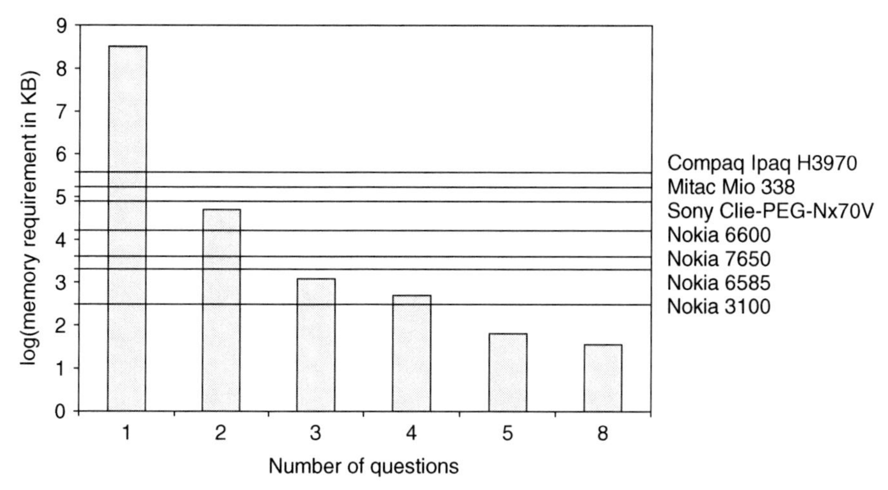

Figure 6.14 The $\langle m, q\rangle$-characterization of the dialog call-flow shown in Figure 6.13.

running the RESEQUENCE algorithm on the call-flow mentioned in Figure 6.13(a) by varying m and finding the corresponding q values.

MASQ

Figure 6.15(a) shows an ideal call-flow corresponding to the call-flow shown in Figure 6.13(a). The ideal call-flow, atomic call-flow, reorganization constraints and the device's memory resources form an input to MASQ. The output of the algorithm is the optimal call-flow with minimum distance between the output and the ideal call-flow. Figure 6.15(b) shows a call-flow for $m = 29$ KB. It requires 12 questions to be answered during the course of the call-flow execution.

The implication of the distance between two call-flows can be observed by comparing the output of RESEQUENCE and MASQ. With reference to the atomic call-flow described in Figure 6.13(a) for $m = 29$, the output of the two algorithms would be different. RESEQUENCE can output a merged grammar corresponding to either 'Flight number' with 'Class of reservation' or 'Class of reservation' with 'Number of seats'. The output is like this because the memory requirement for both the merged grammars is same. So the RESEQUENCE algorithm can arbitrarily pick one of them. However in MASQ, the output call-flow would have a merged grammar of 'Class of reservation' and 'Number of seats'. This is because the resulting call-flow is at the least distance from the ideal reference call-flow as shown in Figure 6.15(b).

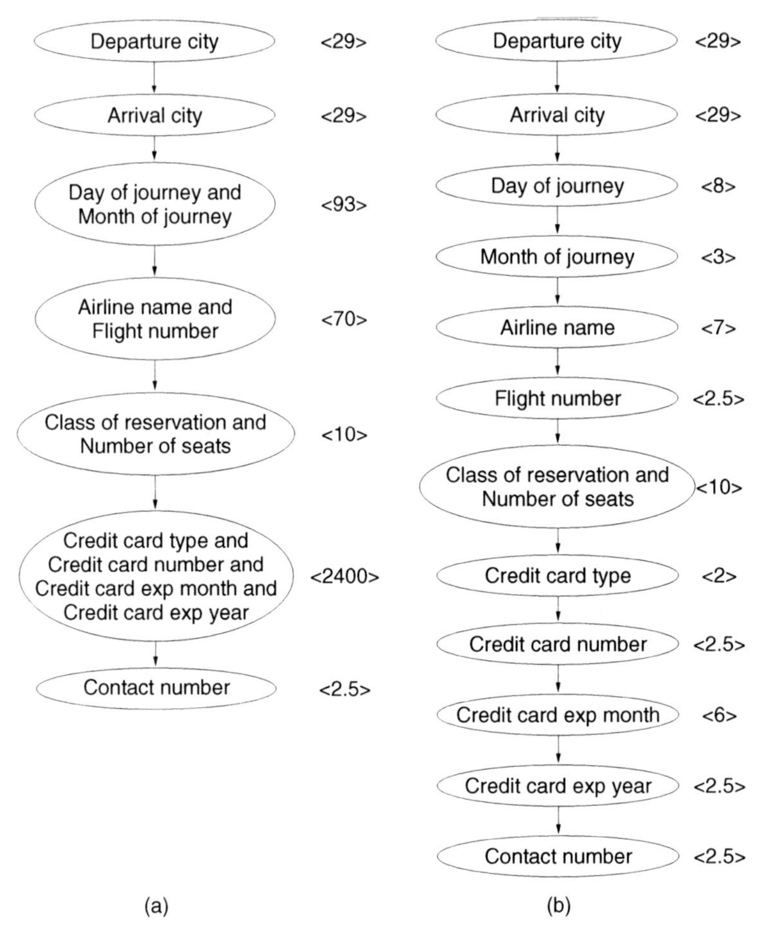

Figure 6.15 (a) An ideal reference call-flow, (b) Output of MASQ *for* $m = 29\,KB$.

We ran an experiment using the reorganized airline reservation dialog call-flow with 14 users. The original call-flow had nine questions and the reorganized call-flow had six. The users were asked to rate the reorganized call-flow on a scale of 1–5 (5 being most satisfactory) for the number of questions, recognition accuracy and dialog completion time. Of the three merges in the reorganized call-flow, one of the merges resulted in poor recognition accuracy and led to decreased user satisfaction. The users were otherwise satisfied with the reorganized dialog.

6.7.8 Splitting dialog call-flows

There is an increasing variety of pervasive devices in use today. More and more applications are being supported on such devices, requiring device-specific application adaptation. We address the problem of speech application adaptation by dialog call-flow reorganization for pervasive devices with different memory constraints. Given a dialog call-flow C and device memory size M, we present deterministic algorithms that alter C to create C_m that fits M by increasing the number of questions and splitting the underlying grammar. We can split a grammar by exposing the intermediate non-terminals in the grammar. The following observation forms the cornerstone of this paper: an and-grammar can be split 'horizontally' and an or-grammar can be split 'vertically' into its components to reduce the memory requirement of the call-flow, at the expense of increasing the number of prompts. We present the algorithm `Minsplit`, and explain its implementation with example call-flows authored in VXML containing SRGS grammars.

A grammar consists of non-terminal symbols which may not have corresponding prompts that are used in user interactions. The essence of splitting lies in exposing these intermediate non-terminals with an explicit prompt. `Minsplit` alters a given call-flow (sequential, tree-type, or hybrid) to fit the memory constraint. It takes as input a VXML dialog containing SRGS grammars and a memory size M. Among other things, it 'exposes' an intermediate non-terminal by converting it to a 'root' node. VXML and SRGS parsers are used to parse the input dialog to build a call-flow. The output of `Minsplit` is converted back to a VXML dialog.

First we examine some preliminaries and describe `Minsplit`. Then we look at the implementation and demonstrate the functioning of `Minsplit` with an example.

Preliminaries

An *and-grammar* of size n is of the form $G := g_1\, g_2 \ldots g_n$, where each component grammar g_i maybe a terminal or a non-terminal. An *or-grammar* of size n is of the form $G := g_1|g_2 \ldots |g_n$. There is a prompt associated with the answer grammar G. Splitting an *and-grammar* G results in 'exposing' two or more g_is by associating a prompt with each. To find an answer for G is equivalent to finding the answers to each g_i and collating them. Splitting an *or-grammar* G results in a regrouping (subset formation) of the g_is into *new* non-terminal symbols G_is. This results in the following grammar: $G := G_1|G_2 \ldots |G_k \text{suchthat} G_i \cap G_j = \emptyset; \forall g_i\ \exists G_j \ni g_i \in G_j \text{and} k < n.$ The idea behind splitting is to *reduce* the number of choices for a prompt at a given level, so that the entire set of choices can fit a given memory constraint. As a result of splitting, at least one additional prompt is generated. Informally, an *and-grammar* is split horizontally into its components, while an *or-grammar* is split vertically into many levels. Thus splitting an *and-grammar* results in new prompts for its components, while splitting an *or-grammar* results in new prompts for each additional level. An *atomic* grammar is one that cannot be split any further. $G := g_1$ and $G := g_1|g_2$, where g_is are terminals. Figures 6.16 and 6.17 show the effects of splitting an *and-grammar* and an *or-grammar* respectively.

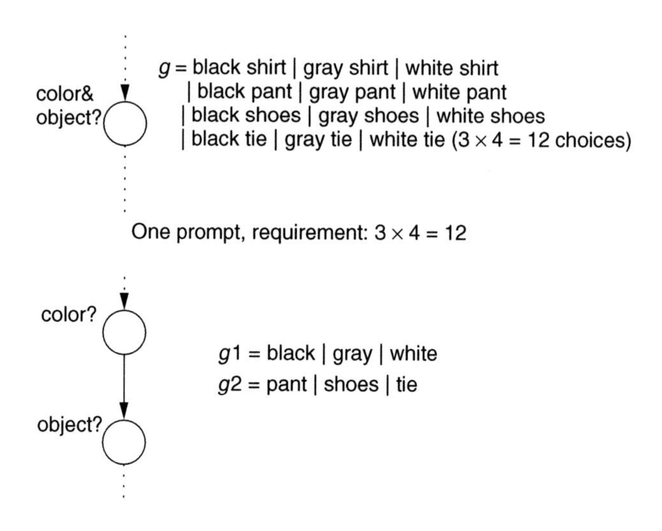

Figure 6.16 Splitting an and-grammar. *A grammar g comprising g_1 and g_2 can be split into two grammars g_1 and g_2 reducing the number of answer choices from $|g_1| \times |g_2|$ to $max(|g_1|, |g_2|)$.*

In the case of an *and-grammar* of size n, splitting introduces a maximum of n new prompts (the old prompt is discarded). The length of the call-flow increases by $(n - 1)$ levels. For an *or-grammar* of size n, splitting may introduce as many as $(n - 2)$ new prompts, and at most $\log(n)$ levels in the tree. Since these operations are all $O(n)$, this ensures that repeated splitting still terminates in polynomial time.

Reorganizational constraints are of two types. One that insists that a certain dialog be split, and a second that forbids a certain dialog from being split. The first set, *must-split*, is a set of dialogs that must be split. The second set, *dont-split*, is a set of dialogs that should not be split. One reason for insisting on splits might be to improve recognition accuracy. One reason for insisting on not splitting might be that it makes logical sense to keep things together, for example credit card number and expiry date. Reorganizational constraints thus provide a mechanism to incorporate various practical considerations and constraints to improve the overall usability and performance of a dialog call-flow.

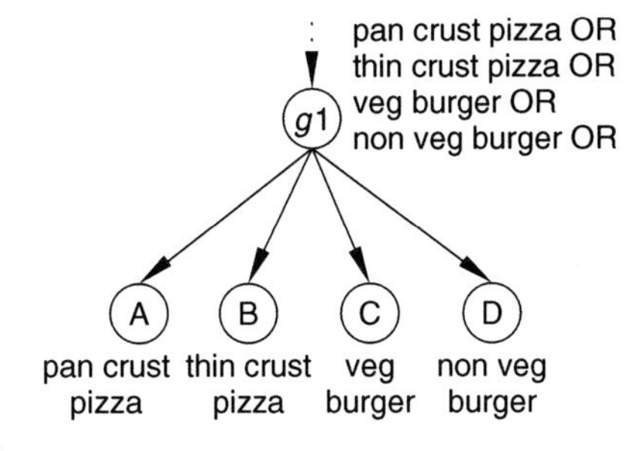

g1 = pan crust pizza | thin crust pizza | veg burger | non veg burger *(2 + 2 = 4 choices)*

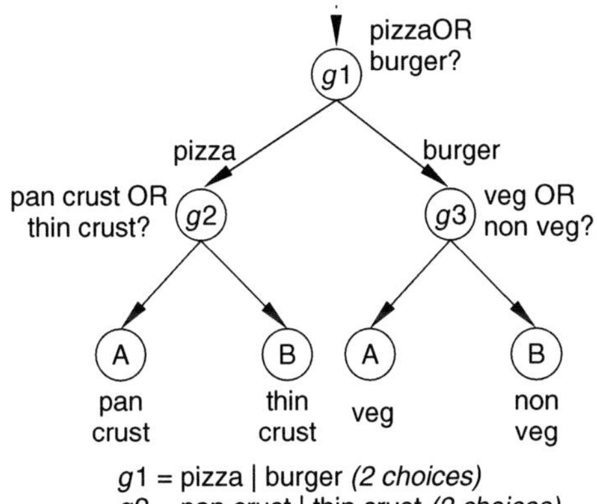

g1 = pizza | burger *(2 choices)*
g2 = pan crust | thin crust *(2 choices)*
g3 = veg | non veg *(2 choices)*

Figure 6.17 Splitting an or-grammar. *A grammar g comprising of g_1 or g_2 or g_3 can be split by introducing intermediate levels, reducing the number of answers choices from $|g_1| + |g_2| + |g_3|$ to $max(g_1, g_2, g_3)$.*

Minsplit

`Minsplit` alters a call-flow C to fit a given memory constraint M within reorganizational constraints. `Minsplit` is described below:

`Minsplit`

1. input: reference sequential call-flow C, memory size M.
2. output: altered sequential call-flow C_m.
3. Construct a graph $G(V, E)$ as follows:
   (a) Represent all dialogs by vertices labeled {1, . . . , n}
   (b) Let $C_m = C$.

   (c) for each vertex $i(1 \leq i \leq n)$

   (d) if !dont-split(g_i) &&
   ($m(g_i) > M$), $split(g_i, C_m)$.

   (e) if must-split(g_i), $split(g_i, C_m)$.

4. output C_m.

5. $split(v, C_m)$

   (a) if (and(v)) // v is an *and-grammar*

   (b) for $(1 \leq i \leq n)$ // v has n components

      i. if $(m(g_i) > M)$ $split(g_i, C_m)$.

      ii. ii. if $(\sum_i m(g_i) > M)$ split at (i-1); else continue.

   (c) else for (all terminals in v) // v is an *or-grammar*

   (d) $S = \lceil \sum_i (m(g_i))/M \rceil$ // S groups

   (e) for (each unexposed non-terminal g_i in v)

   (f) if (non-terminal $m(g_i) > M$) $split(g_i, C_m)$; else expose g_i.

According to step 3(d) and 3(e), split is called when there is a 'must-split' or the memory requirement exceeds M. The split function (step 5) handles both *and-grammars* and *or-grammars*. Step 5(a)–(b) address the *and-grammar*. When an individual component exceeds the memory constraint M, *split* has to be called recursively to break it further. Step 5(b) identifies points where the number of choices exceeds the memory constraint, and splits the grammar at those points. Steps 5(c)–(h) handle *or-grammar*s. In the case of terminals, they have to be accumulated into sets and this is done in step 5(e) while ensuring that the size of any set does not exceed M. Step 5(f)–(h) work on exposing non-terminals, if the memory requirement exceeds M.

Implementation

In this section, we describe the implementation of the call-flow alteration system through grammar splitting. We describe the different types of grammars and show the splitting mechanism for each type of grammar. In the next section we will use a sample doctor's appointment call-flow to illustrate the effects of grammar splitting on the structure of the call-flow.

The grammar splitting is implemented in Java2 (v1.5.0) and the generated corresponding altered dialogs were tested on the IBM WebSphere Voice toolkit, which uses the IBM WebSphere Voice Server for speech recognition and speech synthesis. We tested the system for grammars of several working VXML dialogs and note that it generates syntactically correct dialogs. The implementation works over the SRGS-XML grammar format, but `Minsplit` is independent of the format.

For grammars in the SRGS-XML format, we have used the Java API for XML Processing (JAXP) to build a parser that reads the grammars into a DOM tree. The number of terminals in the DOM tree provide the number of choices of that grammar. This number provides the memory that would be required to process the grammar. In case this memory requirement is greater than the device memory, the grammar has to be split for the call-flow to be accommodated in the device.

An SRGS grammar has a root node which specifies the structure of its children. As was explained in above, grammars are of two types: *and-grammar* and an *or-grammar*. The *and-grammar* has the following structure for a root node:

```
<rule id="rootNode" scope="public">
  <item>
    <ruleref uri="#andPart1"/>
    <ruleref uri="#andPartP2"/>
    <ruleref uri="#andPart3"/>
  </item>
<\rule>
```

There are three rule references defined within the `item` tag. These are the three required items in the grammar. Since all the three items are required, the grammar becomes an *and-grammar*. However, for a *or-grammar*, the root node has the alternative to choose one of the specified rule references. This is represented by the `one-of` tag in the grammar as shown below:

```
<rule id="rootNode" scope="public">
  <one-of>
    <item><ruleref uri="#orPart1"/></item>
    <item><ruleref uri="#orPart2"/></item>
    <item><ruleref uri="#orPart3"/></item>
  </one-of>
</rule>
```

The *or-grammar* shown above expects only one of the items to be said by the user. So the parser understands the grammar structure (and type) by the use of `one-of` and `item` tags in the grammar. The parent of a leaf node in the grammar specifies the terminals of the grammar. The following is an example of terminals in the grammar:

```
<rule id="leafNode" scope="public">
  <one-of>
    <item>Choice A</item>
    <item>Choice B</item>
    <item>Choice C</item>
  </one-of>
</rule>
```

To split an *or-grammar*, each of the rule references within a `one-of` tag can be moved to a separate grammar. However, depending on the memory that a device can support, some of the rule references can be combined in a single grammar. For splitting an *and-grammar*, rule references within the `item` tag and their corresponding definitions are put in a separate grammar. As was the case with *or-grammars*, the number of rule references that are contained in a particular split grammar would depend on the device memory. In case of an *or-grammar*, if the number of choices of a rule reference is too high, then this grammar is split into a number of split grammars, each having some number of choices from the parent grammar.

It is interesting to observe the effect of a grammar split on the structure of the call-flow. A split of an *or-grammar* at a particular node in the call-flow structure results in formation of a tree. The number of children of this node will be equal to the number of grammars into which the original grammar was split. On the other hand, the split of a node having an *and-grammar* will result in a sequential structure at the node. The sequence will however be stretched by the number of grammars into which the original grammar would have been split.

Effect of splitting on call-flows

We illustrate the splitting mechanism for a doctor's appointment call-flow. The call-flow is described by Figure 6.18.

In this example, the grammar used to decode the Date utterance is an *and-grammar*. It captures the month and the day of the month in a single grammar. If the device memory is such that it will not be able to support the grammar since it has a large number of choices (365), then the grammar is split. However, even after splitting, the system needs to know both the values – the day and the month – to schedule an appointment. So the original *and-grammar* `Date`

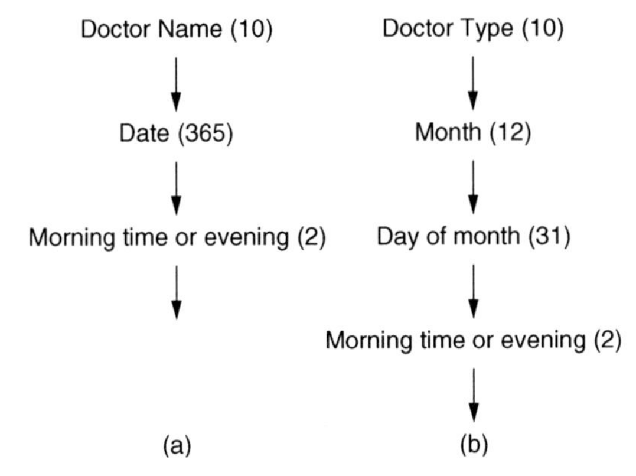

Figure 6.18 A sample doctor appointment call-flow design (a) before split and (b) after splitting of the and-grammar `Date`.

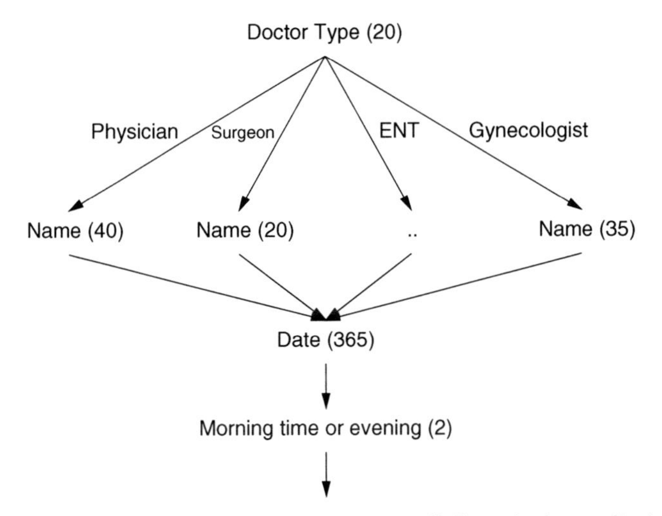

Figure 6.19 A sample doctor appointment call-flow design split in order to reduce the number of doctors in a or-grammar.

is split into two separate grammars. The figure also shows that when an *and-grammar* is split, the call-flow sequence is further stretched by introduction of an additional node in the sequence.

For the example shown in Figure 6.18, the total number of doctors has been set to be 10. Suppose that the total number of doctors were actually 1000. In this case the memory requirement of this grammar would be 1000. In order to split this *or-grammar*, the intermediate node of the type of doctor is used to split the grammar. So the user is first asked about the type of doctor for which an appointment is required. Once the user specifies the choice, then the name grammar for that particular type of doctors is used in the call-flow. This reduces the memory requirement of the call-flow from 1000 to the maximum number of doctors within a particular type. Moreover, it is to be noted that since the original grammar was an *or-grammar*, its split resulted in the introduction of a tree-structured call-flow, even though the initial call-flow was sequential in nature.

6.8 Conclusion

We have addressed the problem of automatically altering a dialog call-flow so that it can meet memory constraints imposed by a pervasive device. The essence of the method lies in splitting a grammar whose memory requirements are larger than the device can support. The reduction in size has to be traded-off with an increase in the number of prompts (questions) in the dialog call-flow. The algorithm `Minsplit` can split both *and-grammar*s and *or-grammar*s.

`Minsplit` has practical implications for automatic dialog call-flow adaptation for pervasive devices. Based on a device profile (characteristics), a call-flow can be automatically adapted for the particular device. We do not address the problem of generating prompts automatically.

More generally, the idea of extending call-flow adaptation for systems that use language models (rather than small 'enumerated' grammars) coupled with an NLU engine would be interesting. Building a mechanism for adaptation in the absence of grammar operations appears to be a very challenging problem.

As speech applications become available on more and more devices, various interesting usability issues are likely to surface. Meeting user expectations without having to manually customize a conversation for every person on every device is a worthy goal for the speech research community.

6.9 Summary and future work

We have introduced, formulated and analyzed device-specific adaptation of dialog call-flows and realized it in the form of SAMVAAD. We believe that the concept of $\langle m, q \rangle$-characterization captures the essence of the adaptability of a call-flow and needs to be probed further for a clearer and quantifiable interpretation.

The reorganization algorithms require an *atomic* call-flow as input. It would be interesting to know how these atomic call-flows can be automatically derived. One potential approach could be to *split* a grammar by exposing the intermediate non-terminals in the grammar. Another objective is the automatic generation of prompts for the merged/split grammars. This might require the use of natural language processing techniques. Our user experiments have shown that we need to take recognition accuracy into account before merging grammars. It would be nice to have a method for estimating the recognition accuracy of merged grammars.

More generally, the idea of extending call-flow adaptation for systems that use language models (rather than small 'enumerated' grammars) coupled with an NLU engine would be interesting. Building a mechanism for adaptation in the absence of grammar operations appears to be a very challenging problem.

As speech applications become available on more and more devices, various interesting usability issues are likely to surface. Meeting user expectations without having to manually customize a conversation for every person on every device is a worthy goal for the speech research community.

Bibliography

3GPP. 3GPP Standard Speech Enabled Services (SES); Distributed Speech Recognition (DSR) extended advanced front-end test sequences. http://www.3gpp.org/ftp/Specs/html-info/26177.htm.

Akolkar, R., Faruquie, T., Huerta, J., Kankar, P., Rajput, N., Raman, T. V., Udupa, R. and Verma, A. (2005) Reusable dialog component framework for rapid voice application development. In *SIGSOFT CBSE*, May 2005.

Banavar, G., Bergman, L. D., Gaeremynck, Y., Soroker, D. and Sussman, J. (2004) Tooling and system support for authoring multi-device applications. *Journal of Systems and Software*, Elsevier Science Inc., NY, USA, 69(3), 227–242.

Braun, E., Hartl, A., Kangasharju, J. and Mühlhäuser, M. (2004) Single authoring for multi-device interfaces. *Adjunct Proceedings of the 8th ERCIM Workshop: User Interfaces For All*, 2004, pp. 4–20.

Comerford, L., Frank, D., Gopalakrishnan, P., Gopinath, R. and Sedivy, J.. The IBM personal speech assistant. In *IEEE ICASSP '01*, pages 1–4, May 2001.

Flinn, J., Park, S. and Satyanarayanan, M. (2002) Balancing Performance, Energy, and Quality in Pervasive Computing, *International Conference on Distributed Computing Systems, (ICDCS) 2002*.

Flinn, J. and Satyanarayanan, M. (1999) Energy-aware adaptation for mobile applications. *Operating Systems Review*, 34, 4863.

Hagen, A., Connors, D. A. and Pellom, B. L. (2003) The analysis and design of architecture systems for speech recognition on modern handheld-computing devices. *ACM CODES+ISSS 2003*.

IBM. WebSphere Voice Toolkit. http://www-306.ibm.com/software/pervasive/voice_toolkit/.

Jameson, A. (1998) Adapting to the user's time and working memory limitations: dew directions of research. *ABIS-98*, FORWISS.

Levin, E. Pieraccini, R. Eckert, W. (2000) A stochastic model of human–machine interaction for learning dialog strategies. *IEEE Transactions on Speech and Audio Processing*, 8(1).

Litman, D. and Pan, S. (1999) Empirically evaluating an adaptable spoken dialogue system. *International Conference on User Modeling*, Banff, Canada, 1999.

Martin-Iglesias, D., Pereiro-Estevan, Y, Garcia-Moral, A. I. Gallardo-Antolin, A. and de Maria, F. D. (2005) Design of a voice-enabled interface for real-time access to stock exchange from a PDA through GPRS. In *Interspeech 2005*, 897–900.

Minker, W., Haiber, U., Heisterkamp, P. and Scheible, S. (2003) Intelligent dialog overcomes speech technology limitations: The SENECa example. In *IUI, Miami, Florida, Jan 2003*, pp. 267–269.

Noble, B. D., Satyanarayanan, M., Narayanan, D., Tilton, J. E., Flinn, J. and Walker, K. R. (1997) Agile application-aware adaptation for mobility. In *Proceedings of the 16th ACM Symposium on Operating System Principles*, 276–287.

Rajput, N., Nanavati, A. A., Kumar, M., Kankar, P. and Dahiya, R., (2005) SAMVAAD: Speech applications made viable for access-anywhere devices. *IEEE International Conference on Wireless and Mobile Computing, Networking and Communications (WiMob2005).*

Rajput, N., Nanavati, A. A., Kumar, A. and Chaudhary, N. Adapting dialog call-flows for pervasive devices. In *9th European Conference on Speech Communication and Technology, Interspeech 2005*, 3413–3416.

Ramababran, T., Sorin, A., McLaughlin, M., Chazan, D., Pearce, D. and Hoory, R. The ETSI extended distributed speech recognition (DSR) standards: client side processing and tonal language recognition evaluation. In *IEEE ICASSP 2004*, 53–56.

Ramaswamy G. N. and Gopalakrishnan, P. S. (1989) Compression of acoustic features for speech recognition in network environments. In *ICASSP98*, 2, 977–980. SPHINX SYSTEM, Kluwer Academic Publishers, Boston, 1989.

SALT. Speech Application Language Tags Forum. http://www.saltforum.org.

SRGS, W3C Recommendation, http://www.w3.org/TR/speech-grammar/

UAProf (1999) WAP-174: UAProf User Agent Profiling Specification. http://www.wapforum.org/what/technical/SPEC-UAProf-19991110.pdf.

VXML, W3C Recommendation, http://www.w3.org/TR/voicexml20/

W3C. VXML W3C Recommendation. World Wide Web, http://www.w3.org/TR/voicexml20/.

Waibel, A. (1996) Interactive translation of conversational speech. *IEEE Computer*, 29, 4148.

Yang, P-C. and Chen, Y-P. (2003) A novel distributed speech recognition platform for wireless mobile devices. In *IEEE Int'l. Conf. on Consumer Electronics*, 358–359.

7

Architecture of mobile speech-based and multimodal dialog systems

Markku Turunen and Jaakko Hakulinen
University of Tampere, Finland

Multimodal interfaces can provide good user interface solutions, but technically they are challenging. Their limited processing power and memory capacities make it hard to include certain functionality, such as speech recognition, into mobile devices. With connected devices, software can use distributed architecture to do part of processing on server computers. In this chapter, we provide examples of such distributed mobile, multimodal applications and take a look at different software architecture models that can be used when such applications are implemented.

7.1 Introduction

The ubiquity of mobile phones has made server-based applications accessible while users are on the move. For example, public transport information services are used while traveling (Turunen *et al.* 2006a). Speech can now also be used to control the phones themselves; many mobile phones include speech technology

---

Speech in Mobile and Pervasive Environments, First Edition.
Nitendra Rajput and Amit A. Nanavati.

in the form of voice dialing and test-to-speech outputs. Speech is introduced since it is useful in many mobile-use scenarios, for example when the user's hands and eyes are occupied while driving a car. Spoken interaction can also compensate for the problems of small or missing displays and keyboards found on typical mobile devices.

High-end mobile phones and PDAs include the possibility to run custom applications, such as midlets in Java-enabled phones). This enables new multimodal applications to be developed, where display, keyboard and voice I/O are used together. Using speech together with graphics, keyboard, and so on in a multimodal manner can make the interaction more efficient and natural since all the available resources of mobile devices can be used in the user interface. Multimodality can prevent errors, bring robustness to the interface, help the user to correct errors or recover from them, and add alternative communication methods to different situations and environments (Cohen and Oviatt 1994) and bring more bandwidth to the communication. For example, Hemsen (2003) has introduced several ways to incorporate speech in multimodal mobile phone applications to overcome the problems associated with small screens. It is also possible to simulate some modalities with other ones, for example in order to enable disabled users (Smith *et al.* 1996). A historical example of multimodality is the famous Put-That-There system (Bolt 1980). In the study by Weimer and Ganapathy (1989), speech inputs made a dramatic improvement in an interface that was based on hand gestures. Similar results have been reported in other studies as well (Hauptman and McAvinney 1993). Different error-prone technologies, such as speech and gesture recognition, can compensate each other, and rather than bringing redundancy to the interface, they reduce the need for error correction (Oviatt 1999).

When talking about mobile interaction, it is noteworthy that the devices introduce new use contexts and new ways of using technology. Because of this, and due to their small physical size, these devices raise new user-interface challenges. The mobile context defines the nature of the interaction: dialogs may be interrupted because the users are concentrating on their primary tasks, or there may be sudden, loud background noises and so on. These interface issues should be taken into account in the design, development and evaluation of such applications. Many of the old solutions of the problems of human-computer interaction are not readily applicable for the small devices that are used while on the move. A combination of speech and other modalities is being applied to this new form of interaction. However, the limited hardware resources in mobile devices complicate building such applications. By distributing the applications between the mobile device and server in a network, new kinds of applications can be built. Such developments raise the importance of software architecture.

From a software architecture point of view, we can separate server-based systems, mobile device-based systems and distributed systems. The conventional telephone-based spoken-dialog systems are server based; all information is processed by the server machines and the mobile phone is used only to transmit audio. This limits the interface since speech and telephone keys are the only available modalities. Systems that are running entirely on mobile terminals (see for example Koskinen and Virtanen (2004) and Hurtig (2006)) can use the display, keyboard, speakers etc. in the device efficiently. However, the limited processing power of these mobile devices limits what is possible in these systems.

By distributing a system between a mobile device and a server, limitations of the other two approaches can be overcome. Since mobile phones also have network connections, albeit with limited transfer rates and high latencies, the applications can also be distributed so that part of the functionality resides on servers, thus enabling features that could not be implemented on the mobile device alone due to processing power and memory limitations. For example, large vocabulary automatic speech recognition is currently beyond the capabilities of small mobile devices. Furthermore, many services and databases may be integral parts of such systems and therefore need to be accessed over the network. Therefore, distributing systems between mobile devices and servers becomes a relevant task. Most current mobile multimodal applications are distributed on some level.

In this chapter, we look at the software architecture challenges of mobile speech-based and multimodal systems. We focus on systems consisting of a single client and ignore multi-device setups such as those studied in certain ubicomp projects, as they have many challenges of their own. First, the common software components are identified. Next, some examples of systems applications are described and analyzed. Generic software architecture models for distributed, mobile multimodal applications are then described and concrete examples given.

7.2 Multimodal architectures

Generic components of a multimodal application and their default relationships are depicted in Figure 7.1. The depicted structure is a call pipeline architecture since it forms a single pipeline where the user input is processed and formed into output. Firstly, input devices capture the user input, and then input parsing forms a higher-level representation of the input. In multimodal applications the input from different modalities is then fused together. Next, the dialog manager of the system decides how the system will respond to the input. In

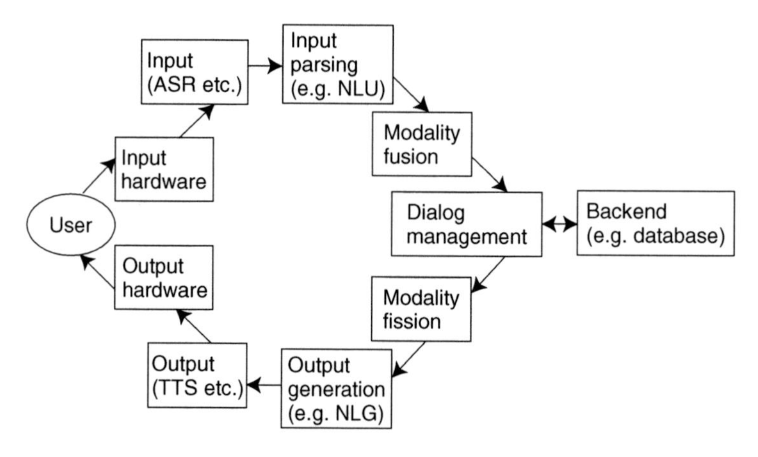

Figure 7.1 Pipeline architecture.

this process, the dialog manager may consult a database or other back-end service. When the output is decided, the modality fission component decides how the message is to be presented using the available modalities. Next, the parts of the actual output are generated and finally the output devices present the output to the user.

In traditional systems, the whole process may have taken place on one actual device, but often individual components have resided on different machines, communicating via a high-speed network. Mobile systems may have everything on the device as well, but connected mobile devices enable distribution. However, the low transfer rates and most importantly the high latency make distribution harder. Data transfer may also be expensive on mobile networks.

While the pipeline model is a straightforward model, this kind of pipes-and-filters model is considered suboptimal for interactive systems (Garlan and Shaw 1993). For example, the response time of the system tends to be slow since every input must travel the entire loop. The model is not very good at supporting concurrent processing by the different modules either. These systems can also be very bulky in mobile and distributed settings.

As said above, in mobile systems we regularly need to distribute the processing between the mobile device and services on the net. Looking at things from the pipeline-architecture point of view, we see that input and output devices are the ones that need to be located on the mobile device since the user is directly interacting with them. This includes not only hardware and software related to screen, keyboard, microphone etc, but also such devices as GPS and cameras.

There are also some components that must be server based. Many applications and information services in particular, need external information sources. For example, a timetable system needs access to a timetable database. Today, more and more applications are modeled using the service-oriented architecture (SOA) approach (Douglas 2003), where loosely coupled and interoperable services are running in networked environments. Many services based on the SOA approach are already available and can readily be used as basis for mobile applications as well. IP-based communication is used between the different devices, and usually XML-based markup-languages are used to exchange information between the components. When multiple services are available, a lookup service may be necessary (Schmandt *et al.* 2004). As an example, the TravelMan application described later in the chapter uses a timetable service, accessed via hypertext transfer protocol and providing timetable information in XML format. Originally built to be used as a back end for a web-based timetable service, the SOA approach enabled the database to be used as part of a multimodal mobile system.

While some components must reside either in a mobile device or on a server, we can see that most of the components can be in either. Most current mobile multimodal speech interfaces are distributed. There are many reasons for this. In some cases, for example in military applications (Daniels and Bell 2001), the entire infrastructure needs to be dynamic to be able to remain usable even when partly destroyed. However, the most important reason is that most mobile devices are not able to run complete multimodal spoken dialog systems because of insufficient resources and missing technology components.

Speech recognition in particular requires significant hardware resources that are not available in typical mobile devices, and thus it is commonly implemented on the server side. Distributed speech recognition – the division of the recognition process into two parts – is one efficient solution to distribute the processing. Typically, the audio processing front end (feature extraction) part of speech recognition takes place in the mobile device, and the actual recognition takes place on the server. This approach has the advantage of reducing the amount of data transmitted over a network, since the audio front end reduces the audio stream to a small set of features. The preprocessing, while requiring reasonable amounts of processing power, requires only a small amount of memory, and can thus take place in the mobile device. The actual recognition process requires more memory and significant processing power and therefore is the part where the power of server computers is most necessary. Industry has worked for a standard solution for such an approach (Pearce 2000).

7.3 Multimodal frameworks

The pipeline architecture provides a generic model, but imposes significant limitations to real-time systems. A pure pipeline is rarely used in real-life frameworks that are used as bases of real applications. One of the best-known architecture frameworks for spoken dialog applications is the Galaxy-II/Communicator architecture (Seneff *et al.* 1998). The difference from the pipeline architecture is that in the middle of the framework is a hub that controls which component receives its turn next. The hub runs a script that can be configured separately for each application to optimize when each component does its processing. The set of different individual components, such as dialog manager, natural language generation etc., can be freely configured into each system as needed. The components can also communicate directly to provide large amounts of data, such as raw audio. The hub passes structured pieces of information called frames between components.

The open agent architecture (Martin *et al.* 1999) is another noteworthy framework. It is not exclusively designed for speech-based or multimodal applications, but has been used in these fields. The architecture calls different components agents, and is built on the idea that the agents communicate with each other. The topography of the network of agents can be tailored for each application.

The Jaspis architecture (Turunen *et al.* 2005a) is a system architecture originally built for spoken dialog applications. Jaspis uses centralized information storage (somewhat similar to Blackboard, as introduced in the Hearsay II system). However, for the selection of components to run, it uses evaluation. Processing is divided into small software components called agents and the best agent is selected for each situation using a set of components called evaluators. The agents and the evaluators are organized into collections, such as the dialog manager. The dynamic selection of an agent for each situation supports more adaptive systems. Jaspis has been used integrally in the Stopman and MUMS systems, which are described later.

The frameworks described can distribute processing onto multiple computers. In practice, each one lets each of the main components reside in a different computers. However, the assumption is that the network is a high speed, low latency one. In mobile systems, the limited network requires more careful analysis of how to distribute the systems between the devices.

7.4 Multimodal mobile applications

This section describes various multimodal mobile applications. For each application, we look at its architecture from the perspective provided by the previous sections. We have participated in the development of each of the systems to some extent and thus can provide technical details and analysis. Each of the systems runs on a mobile phone or a PDA-type device. All the systems are mobile by nature, three providing guidance when traveling using public transport and the first one supporting users with jogging exercises.

7.4.1 Mobile companion

Mobile Companion is part of the larger Companions product range, where these multimodal spoken dialog systems, called Companions, interact with a user in various everyday situations. The Mobile Health and Fitness Companion (Ståhl *et al.* 2008) can be used during physical exercise, such as running or walking, to track the distance, pace, duration and calories burned. The data gathered during an exercise is stored in the device, and can be compared to earlier sessions. Mobile Companion communicates with a stationary home Companion, which provides more generic support for a healthy lifestyle.

Mobile Companion runs on a Windows Mobile device (e.g. a PDA). It provides spoken information for the jogger and a multimodal interface with speech input and output, graphics, together with keyboard and stylus input. All sentences spoken by the Companion via text-to-speech (TTS) are also displayed on the screen as speech bubbles for a graphical agent. The interface is made up of a single screen, as shown in Figure 7.2. At the top of the screen is a status bar that shows the status of the GPS device that is used to track the user's movements during the exercise.

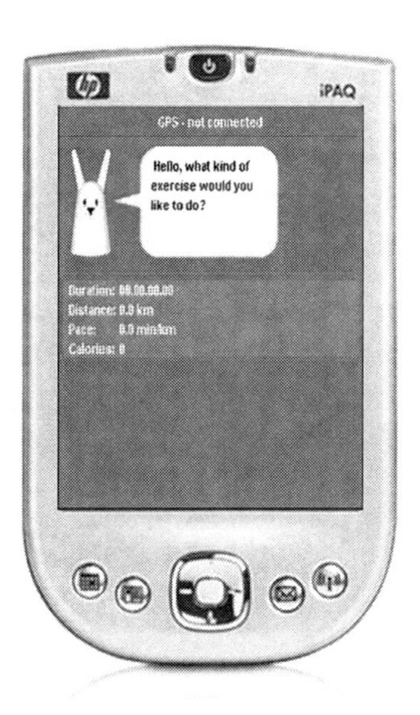

Figure 7.2 Mobile Companion (Ståhl et al. *2008).*

The interaction with the Mobile Health and Fitness Companion is based almost entirely on the use of speech. The mobile prototype is capable of doing TTS and automatic speech recognition (ASR), which means that the normal mode of operation is that the Companion talks to the user via speech output, and the user responds via voice input using a push-to-talk model. Some of the PDA device's keys can be used as shortcuts to input commands, for instance to request the Companion to summarize the status of an ongoing exercise. If the PDA is equipped with a touch-sensitive screen, input to some questions can also be answered by tapping on a list of possible choices on the screen.

The system uses Loquendo's Embedded ASR 7.4 and Embedded TTS 7.4 systems to handle both speech input and output on the mobile device. Since the prototype makes use of both ASR and TTS, it needs a quite powerful processor and also a fair amount of memory. The prototype currently runs on mobile devices such as the Fujitsu Siemens Pocket LOOX T830 (416 Mhz XScale processor, 128 MB RAM) with Windows Mobile 5.

Figure 7.3 shows how the Mobile Companion architecture looks in the model described in the previous section. As can be seen, the system runs completely on the mobile device. Only the back end – the home Companion – is a separate system. Because of this, there is no need for constant access to network. However, as mentioned above, the hardware requirements for the system are rather high. The powerful PDA is capable of running the speech recognizer using medium-sized (hundreds of words) grammar-based language models. The dialog management is a rather straightforward, state transition model.

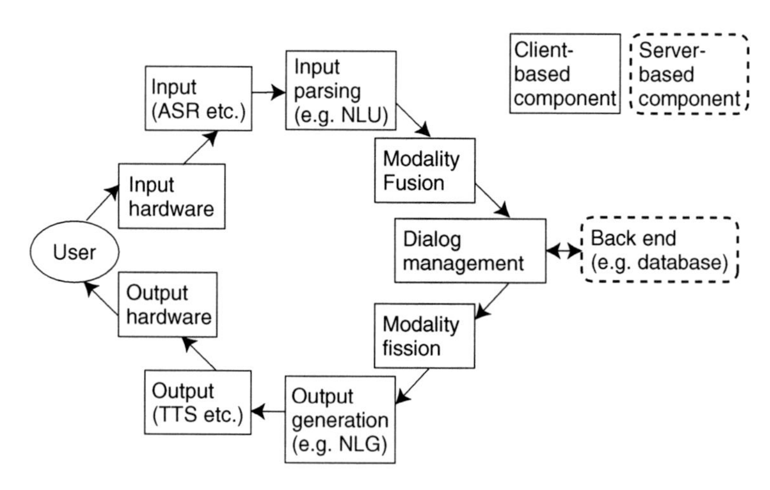

Figure 7.3 Architecture of Mobile Companion.

7.4.2 MUMS

MUMS is a multimodal timetable system running on a PDA. The system can provide timetable information for public transportation and provide navigation instructions for the journey. The interface accepts speech and tactile input, and presents information as a graphical map and using speech. The map on a touch screen interprets all tactile input as locations: a tap denotes a pinpoint location and a circled area a set of possible locations. Speech and touch inputs are recorded simultaneously and time-stamped for later modality fusion. If the user does not provide all necessary information to execute a full route query, the system prompts the user for the missing information. As shown in Example 1, the user can provide a segment of information either by voice or a map gesture.

Example dialog 7.1 (after Hurtig and Jokinen (2006)). The user presents a route query, makes a correction, and finally iterates departure times until a suitable route is found.

U: Uh, how do I get from the Railway station . . . uh.

S: Where was it you wanted to go?

U: Well, there! + <map gesture>

S: Tram 3B leaves Railway Station at 14:40, there is one change. Arrival time at Brahe Street 7 is 14:57.

U: When does the next one go?

S: Bus 23 leaves Railway Station at 14:43, there are no changes. Arrival time at Brahe Street 7 is 15:02.

The MUMS interface runs on a PDA client device and a remote system server. All processing of the user-provided information takes place on the server and, apart from a light-weight speech synthesizer, the PDA functions as a simple client. Fusion of input and fission of output take place in the server among dialog-management and other components. The system is connected to an external routing system and database, which returns route information in XML format. The system is built on the Jaspis architecture (Turunen and Hakulinen 2003).

As can be seen from Figure 7.4, MUMS is architecturally very different to the Mobile Companion system. The mobile device only has the software components related to the input and output devices, namely the code that draws the map on screen, records the stylus input and audio input and plays back audio. The speech synthesizer is also placed on the mobile device in some

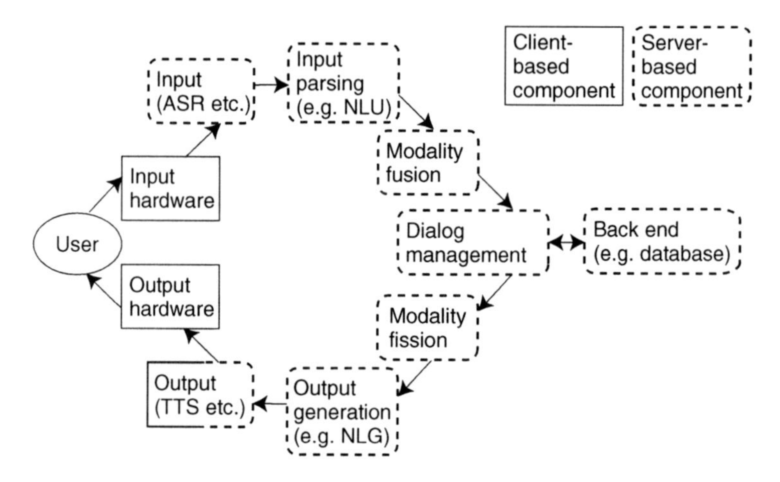

Figure 7.4 Architecture of the MUMS system.

configurations. Such set-up requires constant communication between the client and the server. The system uses an XML-based communication protocol, except for audio, which is sent as binary data. From the server point of view, the mobile system can be seen as a set of input and output devices. However, from the mobile device point of view, it is important to provide instant feedback on the received input, since network communication causes significant delays. The system must inform a user immediately when a speech or gesture has been recorded. This happens within the client UI code and is invisible to the server.

7.4.3 TravelMan

TravelMan is another multimodal mobile application for public transport information. As in the MUMS system, there are two main functions: planning a journey and interactive guidance during the journey. What differentiates the two systems is that TravelMan is designed to be used in mobile phones and thus the interface has been optimized for keyboard input. TravelMan does not have a map but accepts departure and destination locations as addresses. The addresses can be given using speech input, predictive text input optimized for the address data, or using normal multitap text entry. When the user has given the addresses, the system consults a database server on the internet for travel plans and allows the user to browse between the different options. Each route description consists of a number of sub-routes with varying modes of transportation. The lengths of sub-routes are displayed and spoken as temporal distances (durations) between locations, instead of spatial distances. Each sub-route contains detailed information, such as bus or metro stops or streets between the start and end points. After selecting a suitable route, the user can simply listen to how the journey progresses, or navigate in the route description interactively.

The main design principle for the TravelMan interface was to maximize the overall efficiency of the user interface with solutions that work equally well with different modalities and support multiple simultaneous or alternative modalities. Most importantly, the main output modalities, synthesized speech and graphics, support each other. Still, it is possible to use the application with either of these modalities.

While the main idea of the interface is that the active item in the interface is spoken out loud, as shown in Figure 7.5, the spoken content is not necessarily the same as the content on screen. This is because speech and text have different strengths and weaknesses. On the small mobile phone display, it is important to use very few characters for the text. On the other hand, speech, while rather slow as an output, uses full sentences to keep the message easily comprehensible. Several techniques, such as tapering, are used to shorten them. The temporal nature of speech also makes it possible to skip the ends of spoken prompts by quickly moving to the next item similar to Spearcons (Walker *et al.* 2006). Because of this, the system presents the most important information first, which can speed up browsing, especially for blind users. In this way, the same interface is suitable for users ranging from those with normal vision to blind users. In particular, the combination of speech and a graphical fisheye interface supports users with limited vision. Furthermore, this is helpful to all

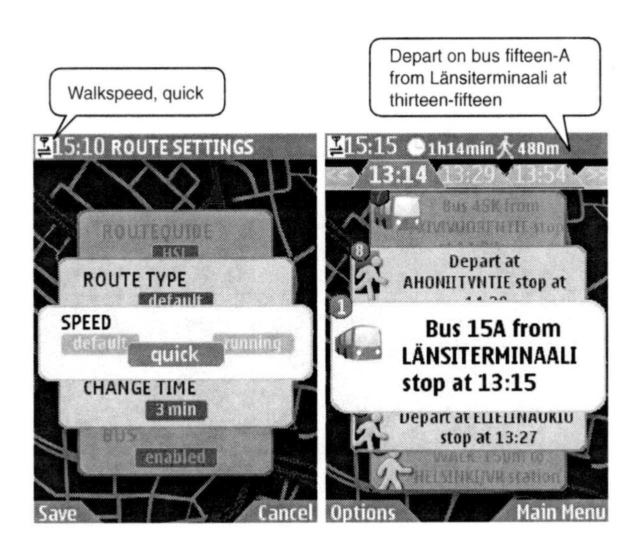

Figure 7.5 TravelMan interface.

users, since in the mobile context we all sometimes have limited vision, for example due to need to monitor our surroundings.

Technology-wise, speech recognition involves a distributed architecture using a server-based Finnish speech recognizer (http://www.lingsoft.fi/). Since the vocabulary is rather large, a server-based system is necessary to make the recognition robust and fast enough. Because the recognition takes place on the server, and as there are better alternatives for menu navigation, it is not efficient in the current interface to use speech recognition for purposes other than giving addresses.

The application runs on MIDP 2.0 compatible Series 60 mobile phones (e.g. Nokia 6600, N72, N95, E61), and uses a locally instaled Finnish speech synthesizer (http://www.bitlips.fi/). A separate Bluetooth GPS device is used for positioning information. The system also utilizes the vibration function of the phones.

As can be seen in Figure 7.6, the TravelMan system is architecturally close to the Mobile Companion. However, the most demanding component computationally – the speech recognition – is server based. This enables the system to work on relatively cheap mobile phones. The speech synthesizer used in the system is a native Symbian application, which limits its usage to smartphones. The system without speech output can actually be run on most MIDP-enabled phones. The way the interface is built requires that speech synthesis be local to the phone. It is used to provide real-time feedback as a user navigates in the interface. Any significant delays would make such use impossible. The fact that the system retrieves the timetable information from the server of a transport services supplier, requires the system to have access to

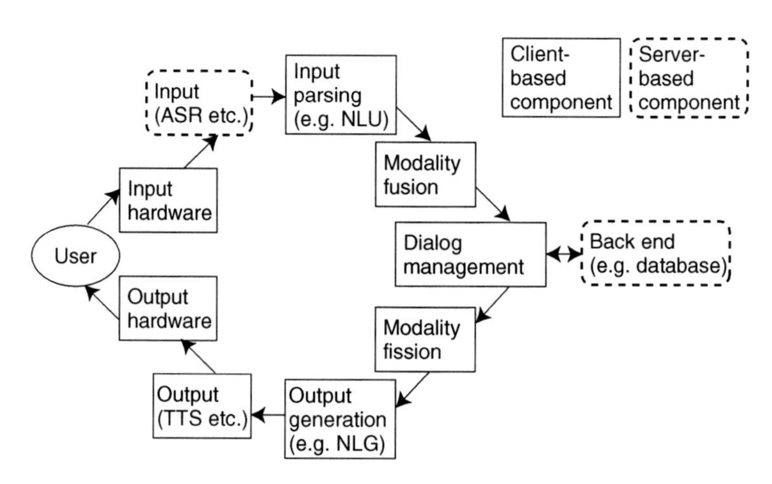

Figure 7.6 TravelMan interface.

the internet. This is a justification for the use of distributed speech recognition in a system, which could otherwise run completely on the mobile device.

7.4.4 Stopman

Stopman is a task-oriented spoken dialog system that provides bus timetables for bus stops. The system offers basic functionality in a system-initiative manner: the system asks for the name of the bus stop while the rest of the functionality, such as browsing the departure times, is available with a user-initiative interface. The spoken dialog version of Stopman was publicly available for four years starting in August 2003 (Turunen *et al.* 2006b). From the records of user experiences, long listings and spoken menus were frequent problems. Users found them boring and complicated, and they easily forgot what was said previously. A distributed multimodal interface for the system was developed to deal with these problems. The solution also provides the richest example of distribution in this chapter.

Figure 7.7 illustrates two interfaces of the second version of the Stopman application: the conventional telephony-based interface on the left-hand side, and the distributed, multimodal interface on a smartphone on the right. The interaction is similar except that the display and keyboard of the phone are used in the multimodal system, while in the telephony interface only speech and touch tones are used to interact with the system.

The multimodal interface uses display, joystick and keypad to support speech inputs and outputs. In addition to the spoken prompts, supporting information is displayed on the screen and menus are presented graphically. The menus are interactive so that users can use speech input or use the telephone keypad for navigation and making selections. Menu items and global options can be selected using the joystick or the keypad just like in the telephony version. When an option is selected it is highlighted and its value is spoken. This kind of multimodality can help users, especially in noisy outdoor conditions where it may be hard to hear system utterances and speech recognition might be unacceptably poor. Graphical information can also help in maintaining the system status, if the user must do something else, like stepping into a bus, while using the system.

As in the TravelMan system, the spoken prompts and the displayed graphical contents are not the same. Once again, information on screen is very brief, whereas more words are needed in order to make the spoken prompts understandable, fluent and less likely to be misheard. However, the information conveyed is the same.

Figure 7.8 shows the architecture model of the Stopman system. Most of the items are split between the mobile device and a server. In the case of speech

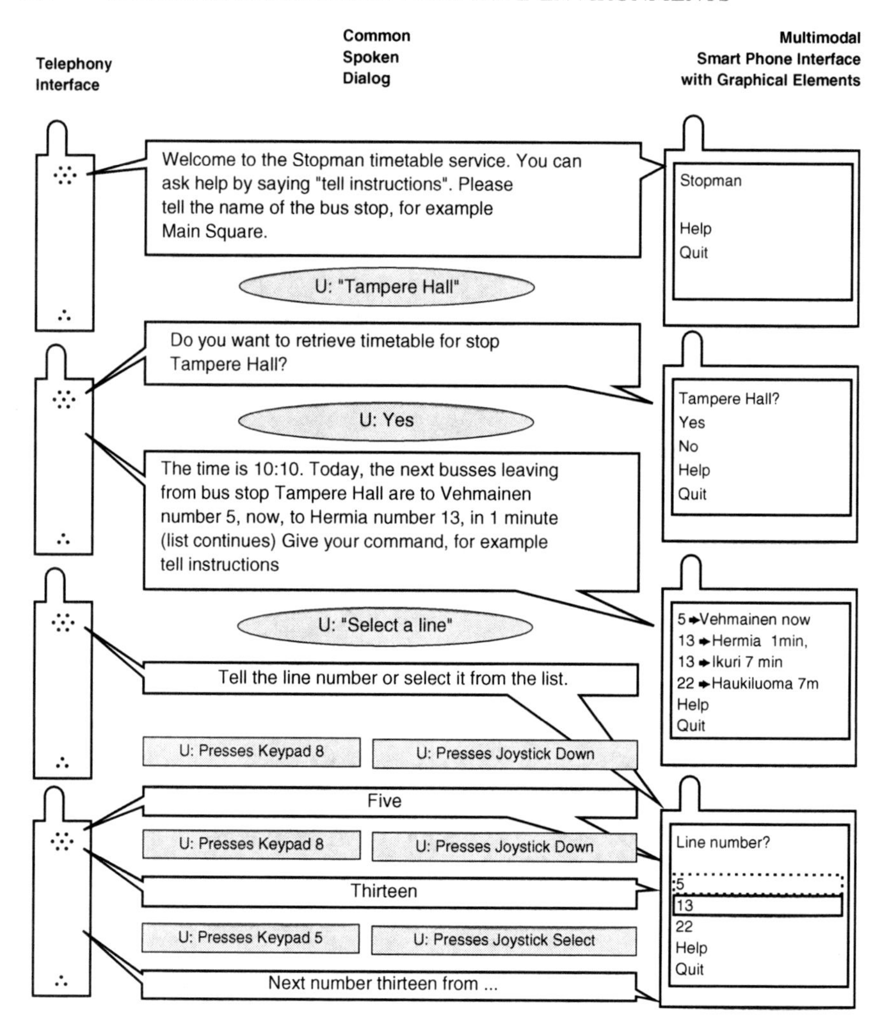

Figure 7.7 An example dialog with the multimodal Stopman system.

recognition and speech synthesis this means that the architecture supports both server-based and mobile device-based components. However, for input parsing, fusion, output fission, and most importantly for dialog management, this means that the processing is indeed distributed. Part of the processing takes place on the server and part on the mobile device. For dialog management this means that overall coordination takes place on the server, while the mobile device handles small tasks such as all the operations related to making a menu choice.

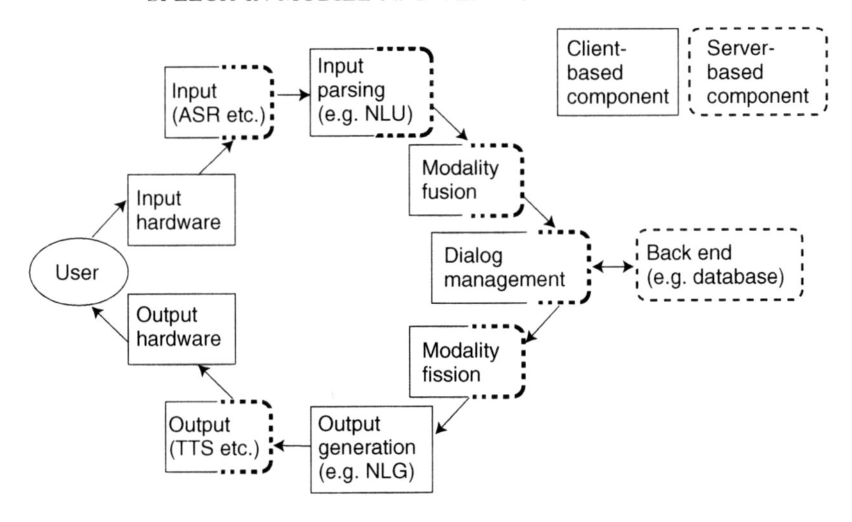

Figure 7.8 Architecture of Stopman system.

The distributed solution enables optimal use of resources. As in the earlier examples, the local device is used to provide immediate feedback such as highlighting menu choices as a user navigates them. Such a feedback loop would be too slow if the processing needed to go to the server and back. What makes Stopman different from the systems described earlier is that the menus and other content in the interface is not hard-coded into the program that is run on the mobile device. Instead, Stopman uses mark-up language to describe menus and other such interaction items.

The Stopman system uses VoiceXML to describe the distributed dialog fragments. Originally, VoiceXML was designed for speech-based interaction. Other modalities can be supported by including additional information in the VoiceXML descriptions as variables or comments. Example 7.2 presents a form with four alternative options taken from the example dialog illustrated in Figure 7.7. The prompt field (line 3) describes what is said to the user, the display variable (line 4) describes what should be presented visually, and the option fields (lines 5–8) describe the possible input options. The prompt and display elements can convey the necessary information without the other being present. This is important, for example, if there is no display or the sound is turned off. Since speech recognition is done with separate components in the server, option fields may or may not cover all the possible inputs: the server-based recognition may have a larger choice of actions than the menu-based interface in the mobile device.

Example dialog 7.2:

```
<form>
 <field>
  <prompt>Do you want timetables for stop Tampere
       Hall?</prompt>
  <var name="display" value="Tampere Hall?" />
  <option>Yes</option>
  <option>No</option>
  <option>Help</option>
  <option>Quit</option>
 </field>
</form>
```

The the example dialog 7.2 shows an example dialog description of the interaction shown in Figure 7.7. It is noteworthy that the dialog description does not tie the interface to any particular device. For example, the dialog presented in this example can be applied to information kiosk interfaces quite straightforwardly.

Depending on the capabilities of the devices used, the distribution of tasks between the server and the mobile device are handled differently by the corresponding dialog interpreter components. This is because the resources of different devices are asymmetrical. For example, one might have a speech synthesizer but no display, while another device has a display but no local TTS engine. The devices must distribute the tasks optimally so that tasks are done where it is possible, and what cannot be done is left for other parts of the distributed system. For example, speech inputs can be recorded in the device and transferred to the server for recognition and further processing. When as much as possible is processed locally, the interface is responsive and we can also use the input and output capabilities of difference devices, such as touch screens and loudspeakers, to their full extent.

7.5 Architectural models

As seen in the example applications, we can split the pipeline architecture from various points, for example having just the back end on a server or using a dummy client (traditional telephone-based system). Anything in between those two extremes, for example having speech recognition and the back end on a server, as in the TravelMan system, requires that the system cannot be a simple pipeline – a local feedback loop must be used to keep the system responsive.

We can also see that in many cases, there cannot be a single-line draws in the the pipeline model to designate the distribution between a mobile device and a server. Instead, individual components can use services or a single component can even be distributed by itself. In the following section, we generalize the previous examples to more generic models.

7.5.1 Client–server systems

In the context of this chapter, client–server systems are those where the client (mobile device) calls one or more servers for specific tasks, such as speech recognition and speech synthesis. Depending on the complexity of the system, the resources required from the mobile device vary. Since it is responsible for the overall control of the system, the more complex the dialog and interaction management, the more resources are needed in the mobile device. The TravelMan system can be considered a client–server solution. The speech recognition and timetable database are the two services the client uses. On a technical level, speech synthesis on the TravelMan system is also a service. It is just one that resides on the same physical device, and thus there is no significant network delay. A similar approach is used in Mobile Companion, where both speech recognition and synthesis are services on the local device and the home system is accessed via the internet. In both cases, the reason for using the local services is that using network socket communication is a straightforward way to enable communication between components written in two different programming languages (Java and native C/C++ programs in both cases).

In the client–server approach, the server-based components can be generic services, modeled after the principles of SOA development. In this way, multiple applications can share, for example, the same speech-recognition service. The code on the mobile client side, however, is very much application specific. TravelMan and MUMS are both examples of client–server-based systems. TravelMan uses a generic timetable service and both systems use basically the same, Jaspis-based speech-recognition server, which can be used as part of server-based systems as well. However, in both systems the number of services used is rather small and a large part of the processing is implemented on the mobile device. This means that most of the implementation work takes place on the mobile platform.

Many client–server architectures include a central hub or a facilitator to connect the resources of the server and the mobile clients together, as seen in Figure 7.9. Such systems usually have most of the processing on the server side. Many mobile multimodal systems have been implemented with this approach. For example, LARRI (Language-based Agent for Retrieval of Repair Information) is a wearable computing system that uses Galaxy-based hub architecture

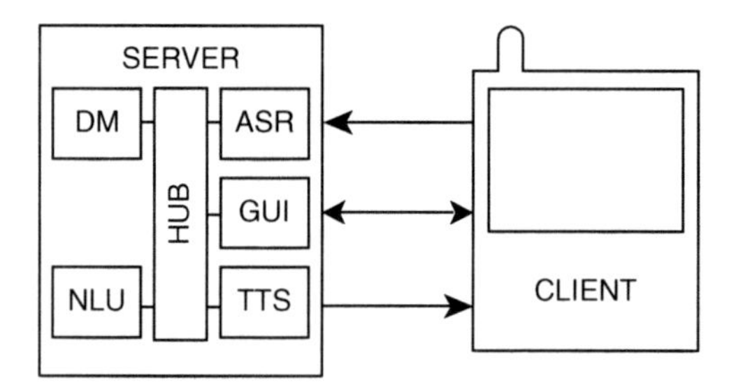

Figure 7.9 Client–server model with hub-based server.

on the server side, and separate components for the GUI and audio I/O on the mobile client side (Bohus and Rudnicky 2002). All resource-intensive tasks, such as dialog management and speech recognition, are performed on the server side. A very similar approach has been used in the MUST guide to Paris (Almeida *et al.* 2002).

7.5.2 Dialog description systems

Optimal use of hardware on a mobile device and also optimal use of the limited network capacity are required to build efficient distributed mobile multimodal applications. Limited hardware forces some processing to take place over a network, and network delays make it impossible to delegate everything to the server. The client–server architecture leaves the task of the distribution to the program written for the mobile device. However, programming these devices tends to be harder than programming servers. To move development work to the server side, and to provide more generic solutions for the architecture problems, description languages can be used. This solution requires a notation for describing dialog tasks with some mark-up and an interpreter (often called a browser) to run the descriptions. This is similar to web architecture, with its HTML files and web browsers. The descriptions are generated on the server side and at least part of them are run on the mobile devices. The Stopman system is an example of such a distribution.

The most widespread dialog mark-up solution is VoiceXML (2006), which has become an industry standard in the area of spoken-dialog systems. The success of VoiceXML itself is one proof of the applicability of markup languages in describing dialog tasks. By default, VoiceXML is designed for server based applications and distribution is used to separate a back-end server from a voice browser server. In a typical VoiceXML setup (Figure 7.10), a mobile device (dumb telephone device in most cases) contacts the VoiceXML browser via telephone network. The browser handles connections to voice services (ASR, TTS)

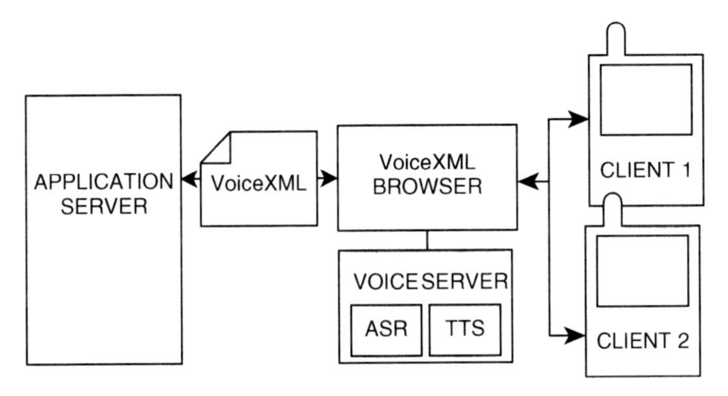

Figure 7.10 VoiceXML architecture.

and back-end applications that reside on the server. Since regular telephones are used by default, the distribution in VoiceXML systems takes place between servers: the back-end server handles the overall dialog and provides VoiceXML description to the browser server, while the VoiceXML browsers handles the dialog descriptions.

In the VoiceXML architecture, all computational resources are located on the server machines. However, we would like to process part of the dialog locally on the mobile devices and use their resources (e.g. TTS, display) to enable more fluent and versatile interaction. Many such solutions are based on mobile clients that are able to run simple dialog tasks generated by the server. In most cases, a markup language such as VoiceXML or SALT (2006) is used for the dialog descriptions. Typically, the mobile device receives a description of a dialog fragment the server has generated, processes it, and provides the results back to the server. This is illustrated in Figure 7.11. In this

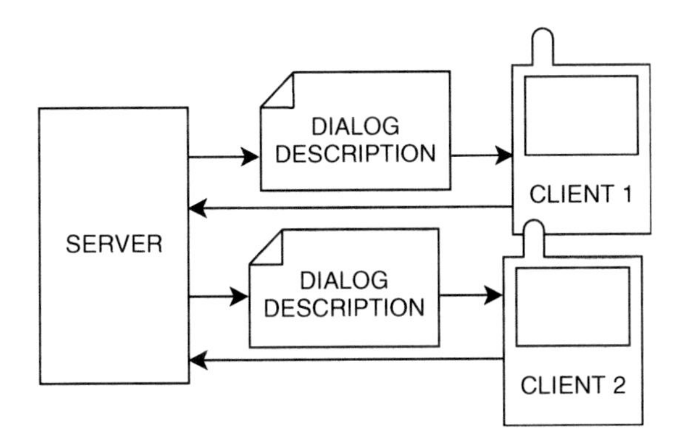

Figure 7.11 Dialog description architecture.

way, similar to web development, only an interpreter for the markup language in question needs to be implemented for each device and implementation of new applications requires programming only on the server side. The markup language may contain program code (such as ECMAScript code in the case of VoiceXML) to be processed locally on the mobile device.

Dynamic generation of VoiceXML-based dialog descriptions are used in various systems (Di Fabbrizzio and Lewis 2004; Pakucs 2002). Recent work with mobile applications includes converting static VoiceXML documents to ECMAScript code to make them compact enough for embedded devices (Bühler and Hamerich 2005). Another approach to manage the limited processing capabilities of the mobile devices is to alter the dialog flow as presented by Rajput *et al.* (2005).

VoiceXML is designed for speech-based systems. There are various markup languages targeted for multimodal interaction. These include XISL (Katsurada *et al.* 2002), SALT (2002), XHTML+Voice (2006) and MML (Rössler *et al.* 2001). They have slightly different approaches, but often the basic principle is the same. For example, XHTML+Voice extends a graphical HTML-based syntax with the voice communication markup taken from the VoiceXML framework, so as to enable multimodal interaction. The distribution of tasks between the server and the client is similar to that of VoiceXML. The generic User Interface Markup Language (UIML 2006) has also been used in mobile multimodal systems (Larsen and Holmes 1998; Simon *et al.* 2004).

7.5.3 Generic model for distributed mobile multimodal speech systems

In order to achieve the maximal flexibility and efficiency in mobile speech applications, the system framework should support distribution of system components, efficient sharing of available resources and distribution of dialog tasks. To enable the distribution of the interaction management, both local and shared resources must be accessible when necessary, and the dialog tasks represented with a common dialog description language that both the server and client devices are able to execute.

Figure 7.12 presents a generic distribution model for mobile systems. The architecture contains dialog management components both on the server and client sides. The dialog management component on the server side generates dialog descriptions containing all the information needed to carry out dialog tasks (e.g. menu selections, correction dialogs, confirmations). The dialog task can be carried out either on the server or the client device, since the client and server use the available resources to actualize the tasks and communicate when necessary. In this way, the resources available in each environment can

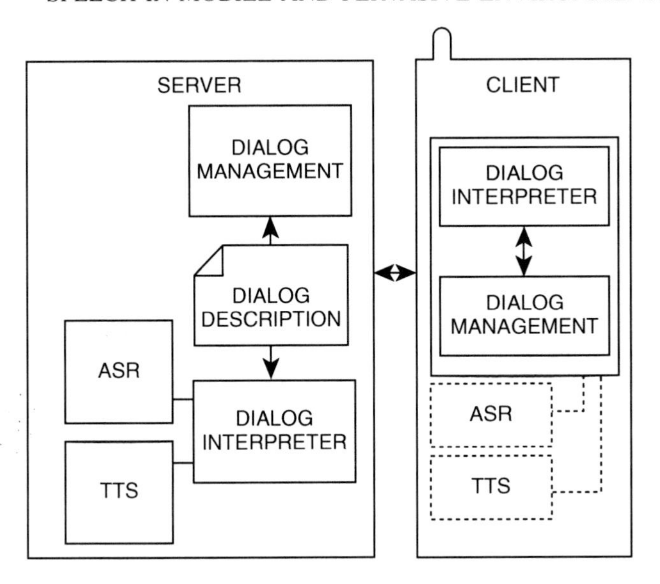

Figure 7.12 Distributed architecture for mobile speech applications.

be used optimally, as devices that know their own resources make the decisions on how the descriptions are actualized. For example, if we do not have a speech recognizer on the mobile device, the dialog interpreters need to communicate in order to share the server-based ASR component. Since both the server and the mobile device are capable of running the dialog description, we can achieve the required level of flexibility.

Markup languages for dialog descriptions, and their dynamic utilization in particular, are crucial in the distribution model described. When we use a markup language to describe the information needed in interaction tasks, we can achieve both the optimal use of resources and consistency. Low-level tasks, for example the selection of optimal output medium, can be distributed to devices that handle the tasks depending on their capabilities. Overall coordination of high-level tasks is kept in the part of the system that generates the descriptions, and complex methods for dialog management can be used. This also enables the system to be distributed physically and makes it possible to change the client devices used, even during a single dialog if needed.

7.6 Distribution in the Stopman system

The Stopman system provides a concrete example of the generic model for distribution. As a final example, we provide a more detailed view of the components used in the system to enable dynamic distribution.

The role of the server in this model is to handle high-level dialog management by generating the extended VoiceXML descriptions described earlier. The mobile device contains an interpreter for executing the VoiceXML descriptions with additional multimodal information. The system also supports the old spoken dialog system interface so that the dialog manager on the server side remains the same and only the technical components (handlers and components under their control) are different in the different setups.

Figure 7.13 depicts the components used to distribute dialog tasks in Stopman. The architecture is based on the generic distribution model (Salonen *et al.* 2005). In this model, the dialog management system (labeled (1) in Figure 7.13) produces the VoiceXML descriptions of dialog tasks (2), such as menus or prompts that should be presented to the user. The distribution of tasks takes place in the communication management subsystem, which is implemented as a modular structure, where the devices are easily changeable without the need to alter the rest of the system. The description generated by the dialog management is handled with the handler devices (3, 4), which use available resources (speech recognition, speech synthesis, display, keys, etc.) to complete the task.

The communication management subsystem contains abstract devices that are used to carry out tasks. These devices represent abstractions of resources that can handle tasks. Devices can also be combined, i.e. they can use other devices to complete the task. For example, a telephone synthesizer is a device that can handle a request to synthesize text and play the result via a telephony card. Separate components, called engines, implement the interfaces to the actual resources such as telephony cards, speech recognizers, synthesizers and communication with smartphones (for the sake of clarity, Figure 7.13 has been simplified and some engines are omitted). I/O agents process the results the devices return, for example language understanding components are implemented as I/O agents. I/O evaluators decide when the communication management system is ready, in other words it determines when to provide results back to the dialog management subsystem. For example, based on the results from one device, an I/O evaluator could determine that it is necessary to run more devices to process the current dialog description.

There are different description handler devices for different terminal devices. In Figure 7.13 there is one telephony device (3) and one smartphone device (4a) on the server side. Both devices are combined devices; in other words they delegate sub-tasks to other devices. The basic functionality of both combined devices is that they get a VoiceXML description as the input, and return a speech recognition result or a menu selection as a result. All other components of the server system, including natural language understanding, dialog management, timetable database, and natural language generation components are the same.

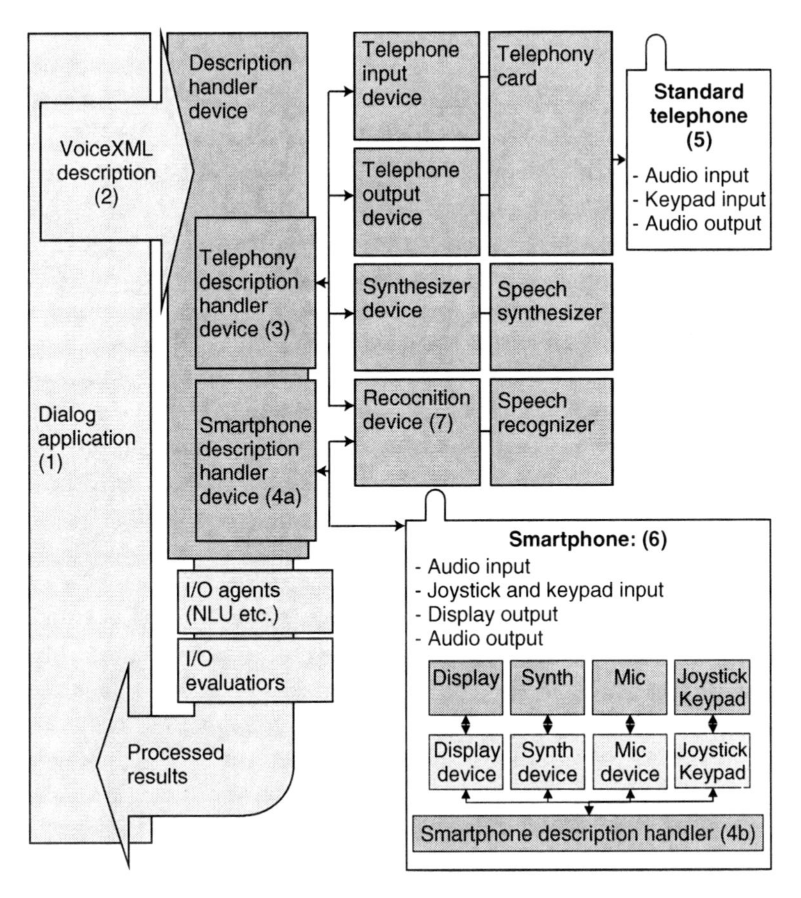

Figure 7.13 Distributed system-level architecture.

In the case of a standard telephony environment (5), all resources (synthesizers, recognizers and telephony cards) are on the server. In the smartphone environment (6), the logic of handling the dialog descriptions is divided between the mobile device and the server. Thus, the combined device (4a) is different from the one in the telephony version (3). Dialog descriptions are handled with components placed on the smartphone, where a local handler (4b) manages the process. As a result, the smartphone returns an audio input or a menu selection. After this, speech recognition is done if needed on the server using the same device as in the telephony version (7). In this way, the I/O agents that perform natural language understanding can be the same in both versions.

As illustrated in Figure 7.13, both the standard telephone device (3) and the smartphone device (4a) are based on the same generic dialog description device.

In this way, the device-dependent and device-independent parts of processing are kept separate, and the system can be easily extended with new devices. At the same time, resources of mobile devices will grow, and some components currently running on the server can be moved to the mobile terminal device. This is possible because of the common system architecture.

7.7 Conclusions

In this chapter, the issues related to the software architectures of mobile multimodal speech applications were discussed. Many successful telephone-based spoken dialog systems have been developed during the last decade. Expanding the interaction to multimodality and making the systems distributed can make the interaction more efficient and natural. However, we face many challenges in this process due to technical limitations and in many cases the system must be distributed between a mobile device and server on a network. The chapter described solutions to the challenges of the distribution. The client–server approach provides a starting point for distributing individual tasks from a system that mainly runs either on a mobile device or a server. A more dynamic and generic model for distribution is available by using a description language generated by a server. Different mobile devices can run parts of the description as well as they can according to the capabilities of the device, and delegate the rest of the processing back to the server. In this way, it is easier to support new devices, as the server part can remain the same and the capabilities of the devices can be used to their full extent.

Bibliography

Almeida, L., Amdal, I., Beires, N., Boualem, M., Boves, L., den Os, E., Filoche, P., Gomes, R., Knudsen, J. E. *et al.* (2002). The MUST guide to Paris: Implementation and expert evaluation of a multimodal tourist guide to Paris. In *Proceedings of the ISCA tutorial and research workshop on Multi-modal dialogue in Mobile environments (IDS2002)*, Kloster Irsee, Germany.

Bolt, R. (1980). 'Put-that-there': Voice and gesture at the graphics interface. *Computer Graphics*, 14(3), 262–270.

Douglas, B. (2003). *Web Services and Service-Oriented Architectures: The Savvy Manager's Guide*. San Francisco: Morgan Kaufmann.

Bühler, D. and Hamerich S. W. (2005) Towards VoiceXML compilation for portable embedded applications in ubiquitous environments. In *Proceedings of Interspeech 2005*, 3397–3400.

Bohus, D. and Rudnicky, A. (2002). LARRI: A language-based maintenance and repair assistant. In *Proceedings of the ISCA tutorial and research workshop on Multi-modal dialogue in Mobile environments* (IDS2002), Kloster Irsee, Germany.

Cohen, P. and Oviatt, S. (1994). The role of voice in human–machine communication. In *Voice Communication Between Humans and Machines*. Roe, D. and Wilpon, J. (editors). National Academy Press, Washington D.C, 34–75.

Daniels, J. J. and Bell, B. (2001). Listen-communicate-show(LCS): spoken language command of agent-based remote information access. In *Proceedings of the First International Conference on Human Language Technology Research*.

Di Fabbrizzio G. and Lewis, C. (2004). Florence: a dialogue manager framework for spoken dialogue systems. In *Proceedings of ICSLP 2004*, 3065–3068.

Garlan, D. and Shaw, M. (1993). An introduction to software architecture. In Ambriola, V. and Tortora, G. (editors), *Advances in Software Engineering and Knowledge Engineering*, Series on Software Engineering and Knowledge Engineering, Vol 2, World Scientific, Singapore, 1–39.

Hauptman, A. and McAvinney, P. (1993). Gestures with speech for graphic manipulation. *International Journal of Man-Machine Studies*, 38, 231–249.

Hemsen, H. (2003). Designing a multimodal dialogue system for mobile phones. In *Proceedings of Nordic Symposium on Multimodal Communications*, 2003.

Hurtig, T. (2006). A mobile multimodal dialogue system for public transportation navigation evaluated. In *Proceedings of MobileHCI* 2006, 251–254.

Hurtig, T. and Jokinen, K. (2005) On multimodal route navigation in PDAs. In *Proc. 2nd Baltic Conference on Human Language Technologies* HLT2005.

Hurtig, T. and Jokinen, K. (2006) Modality fusion in a route navigation system. In *Proc. Workshop on Effective Multimodal Dialogue Interfaces* EMMDI-2006, Sydney, Australia.

Katsurada, K., Nakamura, Y., Yamada, H. and Nitta, T. (2002). XISL: a language for describing multimodal interaction scenarios. In *Proceedings of ISCA Tutorial and Research Workshop on Multi-Modal Dialogue in Mobile Environments* (IDS2002), Kloster Irsee, Germany.

Koskinen, S. and Virtanen, A. (2004). Public transport real time information in personal navigation systems for special user groups. In *Proceedings of 11th World Congress on ITS*, 2004.

Larsen, A. and Holmes, P. D. (1998). An architecture for unified dialogue in distributed object systems. In *Proceedings of TOOLS 26 – Technology of Object-Oriented Languages*.

Martin, D. L., Cheyer, A. J. and Moran, D. B. (1999). The open agent architecture: a frame-work for building distributed software systems. *Applied Artificial Intelligence*. 13, 91–128.

Oviatt, S. (1999). Mutual disambiguation of recognition errors in a multimodal architecture. In *Proceedings of Conference on Human Factors in Computing Systems: CHI'99*. New York: ACM Press, 576–583.

Pakucs, B. (2002). VoiceXML-based dynamic plug and play dialogue management for mobile environments. In *Proceedings of ISCA Tutorial and Research Workshop on Multi-Modal Dialogue in Mobile Environments (IDS2002)*, Kloster Irsee, Germany.

Pearce, David. (2000). Enabling new speech driven services for mobile devices: an overview of the proposed ETSI standard for a distributed speech recognition front-end. In *Proceedings of AVIOS 2000*.

Rajput, N., Nanavati, A. A., Kumar, A. and Chaudhary, N. (2005). Adapting dialog call-flows for pervasive devices. In *Proceedings of Interspeech 2005*, 3413–3416.

Rössler, H., Sienel, J., Wajda, W., Hoffmann, J. and Kostrzewa, M. (2001). Multimodal interaction for mobile environments. In *Proceedings of International Workshop on Information Presentation and Natural Multimodal Dialogue*, Verona, Italy.

Salonen, E.-P., Turunen, M., Hakulinen, J., Helin, L., Prusi, P. and Kainulainen, A. (2005). Distributed dialogue management for smart terminal devices. In *Proceedings of Interspeech 2005*, 849–852.

SALT http://en.wikipedia.org/wiki/Speech_Application_Language_Tags.

Schmandt, C., Lee, K. H., Kim, J. and Ackerman, M. (2004). Impromptu: managing networked audio applications for mobile users. In *Proceedings of MobiSys'04*, 59–69.

Seneff, S., Hurley, E., Lau, R., Pao, C., Schmid, P. and Zue, V. (1998). Galaxy-II: a reference architecture for conversational system development. In *Proceedings of ICSLP98*, 931–934.

Simon, R., Jank, M. and Wegscheider, F. (2004). A generic UIML vocabulary for device- and modality independent user interfaces. In *Proceedings of WWW2004*, 434–435.

Smith, A., Dunaway, J., Demasco, P., Peischl, D. (1996). Multimodal input for computer access and augmentative communication. In *Annual ACM Conference on Assistive Technologies*, 80–85.

Ståhl, O. Gambäck, B., Hansen, P. Turunen M. and Hakulinen, J. (2008) A mobile fitness companion. In *The Fourth International Workshop on Human-Computer Conversation*.

Turunen, M. and Hakulinen, J. (2003) Jaspis – an architecture for supporting distributed spoken dialogues. In *Proceedings of the Eurospeech 2003*, 1913–1916.

Turunen, M., Hurtig, T., Hakulinen, J., Virtanen, A. and Koskinen, S. (2006a). Mobile speech-based and multimodal public transport information services. In *Proceedings of MobileHCI 2006 Workshop on Speech in Mobile and Pervasive Environments*.

Turunen, M., Hakulinen, J. and Kainulainen, A. (2006b). An evaluation of a spoken dialogue system with usability tests and long-term pilot studies: similarities and differences. In *Proceedings of Interspeech 2006*, 1057–1060.

Turunen, M., Hakulinen, J., Räihä, K.-J., Salonen, E.-P., Kainulainen, A. and Prusi, P. (2005a). An architecture and applications for speech-based accessibility systems. *IBM Systems Journal*, 44, 485–504.

Turunen, M., Hakulinen, J., Salonen, E-P., Kainulainen, A. and Helin, L. (2005b). Spoken and multimodal bus timetable systems: design, development and evaluation. In *Proceedings of 10th International Conference on Speech and Computer - SPECOM 2005*, 389–392.

UIML (2006). UIML specification. http://www.uiml.org/specs/uiml3/DraftSpec.htm.

Walker, B. N., Nance, A. and Lindsay, J. (2006) Spearcons: speech-based earcons improve navigation performance in auditory menus. In *Proceedings of the International Conference on Auditory Display, 2006*, 95–98.

Weimer, D. and Ganapathy, S. (1989). A synthetic visual environment with hand gesturing and voice input. In *Proceedings of CHI '89*. ACM Press, New York.

VoiceXML. (2006). VoiceXML specification: http://www.w3c.org/TR/2004/REC-voicexml20-20040316/.

XHTML+Voice. (2006). W3C., XHTML+Voice Profile 1.0, http://www.w3.org/TR/xhtml+voice/.

8

Evaluation of mobile and pervasive speech applications

Markku Turunen[1], Jaakko Hakulinen[1], Nitendra Rajput[2] and Amit A. Nanavati[2]
[1] *University of Tampere, Finland*
[2] *IBM Research, India*

In general, there is no commonly used methodology for evaluation of mobile speech applications. Some methodology is available for evaluation of spoken dialog systems, but mobile speech applications are further behind. In particular, there is much discussion on the possibilities and limitations of field studies. It is commonly stated that laboratory studies are not enough and field studies are necessary. Field studies make us lose most of the control available in laboratory. However, they can reveal many factors we cannot replicate in laboratory or were not aware of. Furthermore, factors such as user experience have become more important, while efficiency-related metrics have become less important. In this chapter we describe how established methods for the evaluation of spoken dialog systems and general methods for field studies, such as contextual enquiry, interviews and diary methods, can be used to evaluate mobile speech applications. Finally, we present several examples of such evaluations.

Speech in Mobile and Pervasive Environments, First Edition.
Nitendra Rajput and Amit A. Nanavati.

8.1 Introduction

Mobile computing is part of our lives as practically everyone has a mobile phone capable of running custom applications, most new laptops can be used everywhere, and various devices in between provide many mobile computing services ranging from games and music to navigation and beyond. The mobility ranges from devices that can be carried around to true mobility, where systems are used while literally on the move. These devices are used in various situations, such as when walking or jogging. Speech is increasingly common in these systems; speech output is important part of car navigation systems and mobile phones commonly enable voice dialing. The hands- and eyes-free interaction that is enabled and the small physical size of the devices make speech useful in these mobile settings. This new mobility alters the way we approach and use information technology. However, mobile use also results in many new problems, ranging from varying weather conditions to a multitude of social situations. This places new demands on the evaluation of the new technology. We cannot get appropriate insight into the usability and performance of mobile computing devices if we test them in the same manner as desktop-based computers.

Both speech-based interaction and mobile use context and hardware provide many new challenges for evaluations of interactive systems. In this article, we give an overview of these challenges and review the currently reported methods which can be applied to the evaluation of mobile speech applications. In particular, we focus on usability evaluation of such systems. Technical evaluation, which has quite different challenges, is not considered in detail. The following sections cover evaluation of speech systems in general, evaluation of mobile systems in general and the effects of the mobile use context. Synthesis of these provides guidelines for usability evaluations of mobile systems. Finally, we present several examples of such evaluations.

8.1.1 Spoken interaction

Speech has distinct features that make it different from, for example, graphical user interfaces. These features are both strengths and reasons for use of speech and limitations and challenges. Speech can be used in mobile systems for many reasons, all linked to these features. Speech may be the only possible modality, it may be the most efficient modality or it may be the most preferred modality. These are strong motives for using speech, since no other modality can outperform it. Speech can also be a supportive, alternative or substitutive modality. The first two cases apply mostly to multimodal interfaces, in which speech can be used in various ways alongside other modalities. In a supportive role speech can be used to achieve more bandwidth and make the interface more robust. As an alternative, speech may be suitable for some users, while other users may prefer other modalities.

Speech is considered a natural way of interaction. It is very common, and usually efficient, in human–human communication. As a symbolic method of communication, speech is also efficient in delivering many kinds of information, especially abstract concepts and relations. However, while speaking to other people is natural, speaking to computers is often considered unnatural. This is emphasized by the fact that computers do not understand completely unrestricted, spontaneous speech. Instead, users must phrase their messages to stay within the language modeled in the system. Furthermore, people have many communication skills and habits that can be difficult to utilize in speech interfaces, but which may cause problems if they are not present in the communication. These include a spontaneous, continuous style of speaking, overlapping and the use of pauses to indicate the structure of the conversation. This dictates that systems must be carefully designed to naturally direct users' speech.

Speech is also an ambiguous form of interaction. Surprisingly many utterances are actually ambiguous and the actual meaning is often reasoned from the context. On a certain level, the ambiguity can be used as an advantage; in most cases it is something to avoid.

Speech is a temporal medium. For an average adult, it is slower to listen to speech than it is to read text. However, speaking is faster than writing. These factors need to be considered when designing user interfaces.

Furthermore, speech is sequential, which limits the output in particular. Everything must be linear, which, together with the slow speeds involved, makes it very hard to present complex structures with speech. Its temporal and transient nature also mean that users' memories are a significant constraint, although it should be remembered that people do not need to memorize everything they hear – this is often misunderstood in speech-interface design and evaluation guidelines.

Speech is also a rather undeterministic signal. Speech production has both interspeaker and within speaker differences and while the sound is traveling in the air, it mixes with other sounds. These factors make speech challenging to recognize. In noisy conditions, people can have a hard time understanding each other, and automatic speech recognition has far greater limitations, making the interaction with a machine error-prone, even in ideal conditions.

Audio, as a medium, is public in its nature. Sounds spread in all directions. Because of this, speech and other audio communications are very useful in making announcements and warnings. They can reach the people present, even if they are not attending to the source of the audio. Audio also leaves users' eyes and hands free, which is very useful in many mobile situations. However, the public nature of audio becomes a problem, when a message is intended only for one person. Privacy-related issues have become common with mobile phones, and speech-based systems must consider the issue.

8.1.2 Mobile-use context

Mobile systems provide many challenges that are not encountered in desktop computing (Holmquist *et al.* 2002). Usage ergonomics are completely different, starting from the small size of the hardware and extending to its physical usage; for example a user may be walking while using a system. Physical movement means that in some cases users may even be physically exhausted. Movement in the environment also exposes users and systems to varying noise levels, lighting conditions, temperatures etc. As the mobile context is often very different from traditional desktop environment, new practices of use may arise as well. Due to the new types of application and usage scenarios, some mobile applications are useful only in certain 'windows of opportunity'. People have different social roles at different times and in different situations, and mobile applications may have to cope with all of these. Context sensitivity therefore becomes a significant feature. A merging of physical and digital realms can also take place, when applications start to provide more spatially aware and related information. As mentioned above, technical limitations, such as network access, which can vary all the time, also affect usage on many levels. It is very hard to know what kind of factors will turn out to be significant.

The effects range from increased noise to complex social situations and increased load due to having to cope with multiple concurrent tasks of the user. Some of the factors can be understood and modeled but much more work is necessary.

The challenges of mobile use can be considered through the concept of context of use. Use context includes such things as the physical conditions like temperature and lighting, the user's physical conditions, such as tiredness or social situation. The biggest challenge is that we cannot know: what the use context will be and what kind of factors in the context are significant (Gorlenki and Merrick 2003). Because of mobility, there are many use contexts and these can change all the time, even during the course of a single interaction with the system. This requires methods that allow designers to understand and study the use context and also new usability paradigms, such as task resumability and integration (Fithian *et al.* 2003).

The mobile-use context results in various factors affecting systems' use and these should be reflected in the evaluation as well. We can identify different effects. Certain variables affect the usage of the system concretely and are reflected in the usability and other measurements. Some variables may affect user satisfaction without affecting the actual usage. For example, users' subjective experience of the system may be better or worse depending on the usage context, even if the objective performance is the same. Some variables can affect users' assessment of their performance on the task they are doing with the system. This can be seen as a subclass of the second class of variables. The final class of factors is those that affect the evaluation itself. We may not be able to use the same methods in context as in a laboratory and even if we can, the results may not be of the same quality. These factors are not distinct classes but a single factor; for example noisy conditions may results in effects in all of the classes.

In a mobile context, users commonly multitask and the usage of the system being evaluated is not the only thing they are interested in. In particular, when mobile computing is used at work, computing is usually a secondary task, supporting the actual work task (Carlsson and Schiele 2007; Gorlenki and Merrick 2003). Therefore, the measurement we are taking may also need to be expanded. We may want the system to cause minimum mental load to the user. In a desktop application this may not be relevant, as long as the load stays within the capabilities of the user population. However, if a user also needs to drive a car at the same time as interacting with their mobile system, the unnecessary load increases the risk of a traffic accident.

Because of the challenges described above, mobile speech-based applications require a wide range of methodologies to support user-centered development and evaluation.

8.1.3 Speech and mobility

Speech technology, in particular, encounters many new challenges in the field of mobile use. In mobile use, the context varies much more than in more traditional computing applications. Different environmental factors affect use. For speech interaction, noisy conditions are a particular challenge. The physical and mental requirements that these different use contexts provide are another factor. From the speech user-interface point of view, this may mean that the user's short-term memory is reduced, thus making providing information with speech even more challenging. It must also be easy to stop using a system for a while and return to it. This means that we must effectively communicate the status of the system to a returning user. Finally, mobile systems must support different social situations. The public nature of speech makes speech interfaces unsuitable for many situations and we may need to think about privacy in the design of speech input and output. Many other complex factors that cannot be foreseen may also arise.

Speech is suitable for mobile computing scenarios because it does not require large physical devices and it is possible to use both speech input and speech output in hands- and eyes-busy situations. At the same time the public nature of speech can be problematic in the varying social situations in which users will use speech interfaces. The tight coupling of speech into the features of mobile computing requires also that the evaluation methods take both factors into consideration.

An issue related to both the evaluation of speech user interfaces and also many mobile solutions is the role of the systems. In contrast to traditional human-computer interaction (HCI) view, where systems are seen as applications or utilities, many spoken-language-dialog systems and mobile solutions can be approached as services. In the peech area, this has been addressed by the application of the SERVQUAL method (Hartikainen *et al.* 2004) as discussed in Section 3.2. Móller and Skowronek (2003), also take this point of view. They use the quality-of-service taxonomy as the basis for their evaluation, specifically in the design of a questionnaire to collect subjective evaluations.

Normally mobile devices have limited hardware capabilities. Processing power, memory, network connectivity, display size etc. are limited. This often dictates the need for distributed solutions, where part of the processing takes place on more efficient server hardware. In speech applications, actual speech recognition may take place on a server, with the mobile device acting only as a front end. Such solutions tend to make the systems slower due to network delays, which may fluctuate with changing network conditions.

8.2 Evaluation of mobile speech-based systems

Because of the features of spoken interaction and mobile use context, the evaluation of mobile speech user interfaces differs from that of traditional computing systems. Perhaps most important difference is the transient nature of speech, which affects many important elements of the system, such as transparency, learnability, error handling, and user control (Larsen 2003b). Mobile use introduces the need to understand the use context. This usually happens by evaluating the system in the field.

In addition to the generally applicable evaluation methods, dedicated and adapted practices exist to evaluate both speech user interfaces and mobile systems. Technology evaluation of speech systems has been done to a great extent because of the probabilistic nature of speech recognition, which dictates the need for evaluation. Mobile systems have likewise raised certain technical issues, mainly as a result of the dynamic network environment.

In the following sections, generic methods of usability studies are discussed first, followed by more specific subjects relevant to the evaluation of mobile and speech-based systems.

8.2.1 User interface evaluation methodology

Studying usability is a long process that lasts throughout the design and development of a system (Nielsen 1994). It may not stop even when the system is ready. We can gain great deal of information by analyzing system usage in real life after it has been deployed. Various methods have been developed to evaluate the usability of interactive systems in general and speech systems in particular.

On a general level, the results from the evaluation can be divided into qualitative and quantitative results – more descriptive information and statistically comparable metrics. Usability metrics, provided by the different methods, are commonly divided into objective and subjective metrics. The former measures users' effectiveness with the system, while the latter measures users' opinions on the system.

On the top level, we can divide these studies into two groups: usability evaluations, which aim at guiding the design into the right directions, and experimental research, which has more academic goals of providing statistically significant results that may be applied on a wider scale.

Another division of the evaluation methodology comprises methods where representative members of the user population use the evaluated software and methods evaluations done by experts using various walkthrough methods and checklist based approaches. When test users are involved, standard usability testing (see for example Nielsen (1994)) is a straightforward option. A user is asked to perform various tasks with the system, the usage is monitored and problems recorded. These tests commonly use a think-aloud protocol (Lewis and Rieman 1993); in other words the user is asked to vocalize their reasoning about the system. In the evaluations questionnaires and interviews can also take place. Questionnaires (Kline 1986) can also be used to gather more generic opinions without exposing a system to the user and to collect opinions after the users have used the system for real. Designing a questionnaire is not a trivial task and, when applicable, ready-made questionnaires (e.g. Hone and Graham (2000)) that have been evaluated and validated provide a safe option.

Data collection can also be distributed to users to take place during actual use. A system can be provided to a user with instructions on how to collect data about their usage. The methods include diaries, traditionally used in psychology but also applicable to studies of technology (Bolger *et al.* 2003). Here users write down notes for example once per day. Collection can also involve photography, audio recording or even collection of physical objects. These other media can be used either alone or together with written diaries or interviews, where they can serve as memory aids (Brandt *et al.* 2007; Palen and Salzman 2002). Automated solutions can also be provided so that a software component asks a user to provide feedback on a system at appropriate times (Isomursu *et al.* 2007).

Use context and requirements for a system can also be gathered by monitoring potential users in real environments. These kinds of contextual inquiry, ethnographic study and field observation have been used since the 1980s in the design of desktop-based systems to understand the actual needs of the users in their work processes.

User opinions can also be gathered using focus groups, where a group of people discuss issues related to an existing or potential system (Krueger and Casey 2000).

When experts evaluate systems without the involvement of users, there are a couple of different methods commonly applied. Experts can do heuristic evaluations, in other words evaluating a system with the help of a list of heuristic rules (Nielsen 1994). The experts use their expertise and apply heuristics to spot problematic designs in the system. Another method is cognitive walkthrough, where the experts consider the system usage from the point of a user's cognition, for example, when users are given a task, the experts consider whether the required functionality can be found from the menu options (Wharton *et al.* 1994).

Various automated methods can also be applied. These often provide large quantity of data relative to the cost of collecting it. However, the data is not as rich as in other methods. Typically, automated methods include setting up a system to log data about its usage and then supplying it to real users for real use. While the data may not be as detailed as in laboratory evaluations, these methods often enable us to collect more realistic data from actual usage in real conditions.

This wide scope of evaluation methods can be explored to find appropriate mechanism for evaluating mobile speech-based systems. Both mobility and speech-based interaction provide their own challenges and limitations for the potential systems.

8.2.2 Technical evaluation of speech-based systems

Because speech recognition is a probabilistic process, evaluation of speech recognizers has been a significant part of the evaluation of speech-based systems. In fact, the evaluation of technical components has been far more extensive than usability evaluation has.

In technical evaluation, when individual components are evaluated, component- specific metrics are used. For speech technology components, the metrics, such as word error rate for speech recognition, are often straightforward to define, since the components can be tested in isolation. The challenge is in collecting and annotating an adequate set of data to test the components.

Measuring the components responsible for interaction management is more challenging because of intercomponent interactions. Danieli and Gerbino (1995) present metrics specifically to evaluate dialog management. Their metrics are transaction success, the calculation of normal and correction turns, and implicit recovery. Implicit recovery is the measure of how well a dialog manager is capable of handling situations where speech recognition or language understanding has failed. In addition, Danieli and Gerbino use a metric called contextual appropriateness, referring to the number of system utterances that are appropriate in the current dialog situation. Appropriateness is related to Grice's maxims; for example, when a system presents incorrect information or too much information, it does not meet contextual appropriateness requirements.

One issue that has been gaining attention in the field of speech user interface evaluation, is what Hirschman and Thompson (1996) call 'adequacy evaluation': the evaluation of the level of performance required of each component in a system. For example, mere recognition rate is not an important figure unless we know what the acceptable level of recognition rate is. Both the users' view of an adequate level and performance of the system can be dependent on factors of mobile usage context, and the context should therefore be considered when the adequacy evaluation is applied. Larsen (2003b) relates adequacy to 'utility' in Nielsen's definition of usability (Nielsen 1994). This issue has been addressed in the SERVQUAL methodology, as adapted to speech user interfaces by Hartikainen *et al.* (2004). In SERVQUAL, participants not only evaluate systems from different points of view but also provide their opinion as to the acceptable and desired standard for each item.

8.2.3 Usability evaluations

Larsen (2003a) claims that 'usability of voice driven services is still poorly understood'. Usability can be seen as a variable we want to evaluate. In such cases usability is usually measured as the sum of some subjective evaluations or it can be based on certain objective measures. However, in quantitative evaluations the measured variables are usually more atomic than the rather complex concept of usability. Naturally, more qualitative usability tests, where usability problems are spotted, are as important in the development of speech interfaces as they are with other systems. Larsen (2003b) argues that improvement in understanding of the subjective measurement of speech-based user interfaces is vital for their development (Larsen 2003a).

In the evaluations of speech-based systems, the Wizard-of-Oz (WOZ) methodology is often applied. In this method, a part of a system is replaced by a human operator. The user is usually not aware of the fact that there is another human in the loop. In this way, systems can be evaluated, even if some part – usually speech recognition – is not yet available or of the required quality.

It is worth noticing that the use of the think-aloud protocol is usually troublesome in the evaluation of speech-based systems. Not only may it not be applicable at all, since some systems would react to the user thinking out loud, but it may also significantly affect the way users end up speaking to the system. In general, when speech-based systems are evaluated, great emphasis must be made on the requirement that test tasks do not put words in the user's mouth. This is already a challenge in graphical applications and is even bigger issue when most of the interaction is based in natural language. Because of this, pictures and pictograms are sometimes used to provide tasks to test users of speech-based systems.

8.2.4 Subjective metrics and objective metrics

Metrics used in evaluations are often divided into subjective and objective metrics. Subjective metrics refer to the opinions of individuals, while objective metrics are'hard figures', which in the case of usability evaluations are derived from measurements of user and system behavior. Normally, subjective metrics are gathered via a questionnaires about test users' opinions. The development of a valid and reliable questionnaire is not a trivial task. There are questionnaires for evaluation of user interfaces, but these have been developed with graphical user interfaces in mind, making their applicability for speech-based and mobile user interfaces questionable. Normally, objective metrics include such numbers as average task completion times and error rates. These are gathered from automatically generated logs or by analyzing video recordings.

For gathering subjective measures in evaluations of speech-based systems, Larsen (2003b) has identified only two questionnaires that have been developed specifically for speech-based user interfaces and that systematically address validity and reliability. These are British Telecommunications Consultative Committee on International Radio (BT-CCIR) and Subjective assessment of speech system interfaces (SASSI). SASSI, which is not recommended for the evaluation of speech-based systems, is a questionnaire that has been developed by testing a pool of 50 questions. Analysis based on 214 completed questionnaires resulted in a set of 34 questions, which were grouped by six user perception factors. Hone and Graham (2000) call these system response accuracy, likeability, cognitive demand, annoyance, habitability and speed. Hone and Graham consider SASSI to 'have face validity, and a reasonable level of statistical reliability'. SASSI is focused on speech-based interaction and does not take mobile context into account. Its applicability has not been specifically tested with the effects of mobile context in mind.

Currently, there are no specific questionnaires for evaluating mobile systems in general. Such a questionnaire would otherwise have been widely accepted.

In the evaluation of spoken-language-dialog systems, the most widely used objective metrics include:

- the percentages of correct system answers, successful transactions, repair utterances, sentences containing more than one word, and user-initiated turns;
- the number of "help' requests and barge-ins, completed tasks, and subtasks;
- dialog and task completion times, mean user and system response times and the mean length of utterances (Larsen and Holmes 2003a).

Surveying the literature, Mǿller (2005) reports 36 different objective metrics, which he divides into five categories: dialog- and communication-related, metacommunication-related, cooperativeness-related, task-related and speech-input-related. Many of these metrics measure the entire system or interaction. As can be seen, many of these metrics are closely related to technical components and their adequacy. Subjective and objective metrics can also be combined into a single model. As an example, PARADISE (Walker *et al.* 1997) provides an entire framework for conducting studies of spoken-language-dialog applications. It separates task requirements from dialog behaviors and thus allows comparison of different applications. During the experiments, both subjective and objective metrics are collected. A regression model to relate users' subjective measurements of satisfaction (which are taken to model usability) and objective evaluations (task success and costs), is calculated. The model shows how much each element of the objective measure of the system performance affects user satisfaction. Möller *et al.* (Mǿller 2002, 2005; Mǿller and Skowronek 2003) have also experimented with other methods beyond correlation models within a similar framework.

There has been discussion on the use and meaning of subjective metrics. For example, in PARADISE-based studies, subjective metrics that are used to evaluate user satisfaction usually consist of a short questionnaire with

Likert-scale answers; user satisfaction is a sum or average (Smeele and Waals 2003) of the numeric codes of the answers. It has been questioned whether a single number derived in such a manner is indeed a relevant measure of user satisfaction, let alone usability. Móller (2002) states that a simple arithmetic mean cannot be used as a measure of user satisfaction.

Another issue raised in these studies is the difference between the task completion rates as perceived by a user and as observed by the test conductors. These two judgments have been found to differ (Walker *et al.* 1998), and some researchers (Móller 2005) use both. In some cases, user perception of task success has been used instead of the kappa coefficient in PARADISE applications (Walker *et al.* 2000). This is questionable since in this case users' subjective evaluations can be found from both sides of the regression model.

8.2.5 Laboratory and field studies

Because of the need to understand use context, field studies are often advocated for the evaluation of mobile systems. By taking the evaluation into the field, we can discover many issues that would not be encountered in the laboratory.

Despite the recommendations of the use of field studies, Kjeldskov and Graham (2003) found that laboratory studies are the most common way to evaluate mobile systems. Applied research is even more common. Context is rarely studied. Lately signs of more field studies, recent HCI literature, for example the review of latest chi, while not all, several field studies (Brewster *et al.* 2007; Jones *et al.* 2007; Mákelá *et al.* 2007; O'Hara *et al.* 2007; Salovaara 2007; Seager and Fraser 2007; White *et al.* 2007;).

Since the evaluation techniques for mobile systems are still in a state of rapid development, it is natural that not much standard methodology has matured. For example, there are no widely agreed questionnaires for measuring users' attitudes towards systems and services.

It has been found that laboratory and field tests provide different results in usability evaluations. For example, in studies by Duh *et al.* (2006) more usability problems were found in the field, especially critical problems (caused by noise levels, moving-train settings, lack of privacy, more required effort and additional stress). Some problems could be found only in field. On the other hand, Kjeldskov and Stage (2004) found fewer usability problems in the field because of their focus on the think-aloud protocol.

The additional stress in the field resulted less think-aloud and therefore less material to detect problems from. Brewster *et al.* (2002; 2007) found similar but much weaker results in a quantitative field study compared to the laboratory. They strongly advocate the need for field tests. Kaikkonen *et al.* (2005) and Betiol and de Abreu Cybis (2003) did not find significant differences between the results of laboratory and field evaluations. Barnarda *et al.* (2005)

in their extensive tests found a few more problems when users were walking outside instead of walking on a treadmill in a laboratory, but the difference was not considered great. However, walking outside was seen to provide a more realistic user experience. Similarly, Baillie and Schatz (2005) did not find significant differences in actual usability problem detection rates, but user attitudes were found to differ. Interestingly, actual usage conditions can make users more favorably disposed toward a system. Heuristic evaluation can also be undertaken in the field.

In their experiment, Po *et al.* (2004) found scenarios that would provide most of the context related information in the laboratory, but some different physical-hardware-related problems in particular were encountered when evaluating in real conditions. However, the additional findings were not considered worth the cost.

Looking at the studies mentioned above, different researchers have had very different experiences when comparing laboratory and field studies. Certain initial generalizations can be made, however. When evaluations were of more traditional GUI-based interactions, such as a set of WAP pages, mobile evaluation did not provide significant new data. Where field studies seem to make a difference, is where there is more physical interaction: the physical design of the device (such as displays that are too dim) or the use of new modalities (such as voice) where noise levels and social situations are more significant.

Field tests have problems (Kjeldskov and Stage 2004): it is complicated to set up realistic studies, hard to set up evaluation techniques (think aloud etc. or video taping) (Weiss 2002) and limited control complicates data collection.

8.2.6 Simulating mobility in the laboratory

Physical movement and language are mostly, but not completely, separate parts of human information processing (Shneiderman 2000) and we are able to speak and walk at the same time. However, when more logical reasoning is needed, divided attention becomes an issue.

One attempt to tackle the effects of increased cognitive and physical load on users in evaluations has been made by Kjeldskov and Stage (2004). They provide two models for field conditions. The first one covers the physical movement and attention needed. They conduct tests in the laboratory with various physical activities in addition to the actual task on the application being tested. The conditions include a case where a user is sitting, another where the user is walking on a treadmill at constant speed and yet another where the walking speed is varying. Finally, the user can be asked to walk on a changing course, either at constant or at varying speed. In their evaluations of these methods, a control condition was a test conducted walking outdoors on a real street. Another model by Kjeldskov and Stage (2004) tries to model the divided attention required when using systems in a mobile setting. To do this, users were asked to use a system while simultaneously playing a physical game. The methods model some of the factors that occur in real-life usage, while still enabling the tests to be performed in the laboratory, thus keeping the data collection and other practical arrangements manageable.

Other factors one may try to simulate in the laboratory include lighting conditions, especially darkness, noise, and possibly even temperature. Technical issues, such as network coverage may also be simulated.

8.2.7 Studying social context

A completely different factor, which is very hard to model in laboratory studies, is social context. With current knowledge, it is best studied with intervention studies (e.g. Salovaara (2007)), diary techniques (e.g. O'Hara *et al.* (2007)), automatic data collection, and interviews during free use (and possibly with ethnomethodology).[1] These are not methodologically trivial. For example, participatory research versus interviewing are different; do we want to affect users or not? (Salovaara 2007).

Many of these field study methods can be applied in more phases of the system development than simply during the evaluation of a prototype or a system that is ready for launch. In the initial design and in the requirements-gathering phase, field studies, such as contextual inquiry and diary techniques, can provide invaluable data. This information can guide our design in the right direction. This approach has been acknowledged already in the design of desktop applications. In mobile applications, where usage context is much more varied, it is even more important. However, it is also much harder, since we may not know beforehand all the usage contexts that will arise. Understanding the usage context from the very beginning can also help us to better evaluate the usability when a functional prototype is ready. After a system has been implemented and given to users, field studies can be carried out using log-file data collection. While this only provides us certain abstract view of what kind of interaction takes place and we can only guess what the usage context and users' goals were, we get large amounts of data that reports the actual usage. Weiss (2002) considers log analysis a very powerful tool.

8.2.8 Long- and short-term studies

Most laboratory evaluations are short term; a user is introduced to the system and all the measurements are done during one session that usually lasts about one hour. Long-term studies, where data is collected over several months,

[1] Ethnomethodology is an ethnographic approach to sociological enquiry.

thus providing information on how users learn to use a system and how their relationship with it develops, can also be made. In the laboratory, long-term studies are reasonable only to assess the effects of learning. However, in the field longer-term evaluations are more common because they let users experience natural usage situations for the evaluated system and let systems be exposed to a variety of natural usage contexts. These studies can be cross-sectional, with a random sample of users every time the data is gathered, or longitudinal, with data gathered from the same set of users several times so that the effects of long-term usage can be studied.

We have studied a mobile spoken language dialog system (SLDS) both with more traditional usability methods and through analysis of logs and recordings from an extended (30+ months) period of real use. We found that after a month or two the data became different from usability-test-type data, with short- and long-term usage being rather different (Turunen *et al.* 2006). In mobile use, the initial calls are often made from a more favorable environment, e.g. from an office. However, when a potential user has evaluated the system and found it to be potentially helpful, usage moves to real mobile situations, where new kinds of usage may emerge. In real-life use, only some functions are used, unlike usability studies which typically force participants to use a wide range of functionality.

8.2.9 Validity

Field studies are usually (in almost all fields of science) qualitative. The problem with field tests is that there are many variables that cannot be controlled. This makes the tests very weak; we get lot of variability from all these variables. Because of this it is very hard to make a statistical evaluation of field tests. However, we can make qualitative evaluations and try to identify various significant variables and, as necessary, include them in laboratory studies. It is also worth noticing that long term field studies with automatic data collection can provide such large amounts of data that the inherently great variability can be overcome to a great extent. Such tests also tend to have great ecological validity – realistic use conditions, as discussed briefly above – but may have issues with unrepresentative user populations.

Variables introduced by the field tests may affect not only usability and the metrics representing it (or whatever we are measuring). They also have an effect on the organization of the tests; we may not be able to apply the same techniques as in the laboratory.

For example, use of a talk-aloud protocol can lead to many challenges, starting with the noisy conditions and varied social situations that users will experience. These factors make the talk-aloud approach unacceptable, since users will have such an increased mental load that it becomes hard for them to speak (Kjeldskov and Stage 2004). Mobile factors may also affect users' expectations of the system (cf. adequacy evaluation) and their evaluation of their task completion.

It is also possible that the test setup we use in field tests generates effects that would never be experienced in real usage situations. Just by moving tests from a laboratory to the field we do not necessarily increase the ecological validity of our tests – the validity may even decrease. While tests take place in the field, the test setup still has its effects and those effects may be harder to understand than those of a laboratory test.

Therefore, the design of a test setup so as to make the test valid is just as important in field studies as it is in the laboratory.

To restate the dilemma of choosing between laboratory studies and field studies: field studies search for greater ecological validity, possibly with a cost in terms of the tests' internal validity. Varying definitions of different kinds of validity can be found in the literature. When discussing the external validity of evaluations, we follow the definition of Bracht and Glass (1968):

> Two classes of threats to external validity are identified: (1) those dealing with generalizations to populations of persons: 'population validity'; and (2) those dealing with the 'environment' of the experiment: 'ecological validity'.

While external validity deals with the ability of experimental results to generalize to the 'real-world' population, ecological validity is achieved when the experimental procedures resemble real-world conditions. While these forms of validity are closely related, they are independent – a study may possess external validity but not ecological validity and vice versa. (i.e. generalization of results and similarity of settings). In virtually all studies there is a trade-off between experimental control and ecological validity.

Internal, external and especially ecological validity are the key here. Laboratory tests are not ecologically valid, field tests can easily be internally invalid. The idea of laboratory tests is to control all the variables that complicate the making of measurements in the field. However, in mobile system evaluation, many of these variables may be very significant. The problem is, as mobile applications are still emerging, we do not know which variables are significant, and should therefore be controlled for in the laboratory (Kjeldskov and Stage 2004). Therefore, field tests must be part of our collection of methods, especially in usability and similar studies, where we try to understand actual use in all its relevant aspects.

One might consider external validity to always, or at least usually, be better in field studies. However, if there is no internal validity, there is nothing to generalize and validity depends on what we are generalizing. For example, we can pick a small group of participants in the field and try to generalize to a different group of people. But because of the low internal validity, we easily lose external validity in the field (Anderson and Bushman 1997).

From the discussion above we can draw some conclusions. First of all, field tests are considered very important. Secondly, how field tests of any particular variable should be conducted is not clear. To formalize this idea, field tests include many uncontrolled variables, which are controlled in laboratory studies. The idea of a laboratory test is to control all the variables we do not want to vary. But we do not know all the variables that affect the use of a mobile system in real environments and therefore we cannot evaluate them in a laboratory, since we could easily have some very significant variable wrong. Because of this, we must conduct field tests to have these unknown variables adopt their real values. It may also be that we know some of the variables but are not capable of setting up appropriate tests for them in a laboratory. Social context is one such variable.

8.3 Case studies

We have developed several systems to provide timetable information over the years (Turunen *et al.* 2005, 2007), and conducted numerous evaluations for them (Melto *et al.* 2008; Turunen *et al.* 2006). In this section we present two case studies from applications targeted for mobile use. In these studies, both laboratory tests and public pilots were used to evaluate the applications. We present these evaluations focusing on the procedure rather than results, in order to demonstrate how multimodal mobile speech systems can be evaluated.

In both cases we show how the results are affected by the usage conditions. First, we present the study of the Stopman bus timetable system. Second, we present the study of the multimodal TravelMan public transport guidance systems.

8.3.1 STOPMAN evaluation

The Stopman system provides timetables for each of the about 1200 bus stops in Tampere City area. The aim of the system is to satisfy busy users with the first timetable listing, as demonstrated in Example 1. This is crucial, since when people are on the move and need to know when and where to catch a bus, they require simple functionality and efficient, robust interaction without long dialogs. This is quite different from the typical web-based timetable query made in front of a computer, and is a good an example of how the mobile context influences the interaction. Robustness is another key feature, since lessons learned from previous spoken dialog timetable systems suggest that open, user-initiative dialog strategies based on data collected from human–human interactions fail to provide sufficiently robust interfaces for real use, as seen in our own (Turunen *et al.* 2005) and similar studies (Ai *et al.* 2007).

In the case of the Stopman system we developed a task-oriented interface that provided the basic functionality in a system-initiative manner, while the rest of the functionality was available through a user-initiative interface, thus creating an efficient and reliable interface for busy mobile users and more functionality for other users.

Example in Table 8.1 demonstrates Stopman usage. At the beginning of the call, the system requests the user to give a bus stop name (S1). The most fundamental information is included in the initial timetable listing, which explains the length of the response shown in (S3). After this, the rest of the functions are available. Available functionality includes the ability to navigate through the timetable, selection of specific bus lines and specification of a certain time (U3).

Table 8.1 An example dialog from the Stopman system.

S1:	Welcome to Stopman. You can ask help by saying 'tell instructions'. Please tell the name of the bus stop, for example 'Central Square'.
U1:	*'Alexander Church'.*
S2:	Do you want to retrieve timetable for stop 'Alexander Church'?
U2:	*'Yes.'*
S3:	The time is 10:10. Today, the next buses leaving from bus stop 'Alexander Church' are to 'Vehmainen' number 5, now (list continues)... Please give your command, for example 'tell instructions'.
U3:	*'Select line.'*
S4:	Please tell the line number. You can list the lines by saying 'list lines'.
U4:	*'Thirteen.'*
S5:	Next number thirteen buses from 'Alexander Church' are to Hermia in 1 minute, to 'Ikuri', in (list continues)...

In order to evaluate the Stopman system we divided the interaction into ten types of user inputs (Table 8.2), and studied how their use changes over time and under different evaluation conditions, as described below.

Table 8.2 Stopman functionality categories.

	Description	Example
1	Mandatory functionality	*'Main library'*
2	End of call	*'Thanks, goodbye!'*
3	Help requests	*'Tell instructions'*
4	Repeat requests	*'Repeat the last one'*
5	Confirmations	*'Yes'*
6	Advanced functionality	*'Select another day'*
7	ASR rejections	⟨NOT RECOGNIZED⟩
8	Missing inputs	⟨SILENCE⟩
9	Invalid inputs	⟨INVALID DTMF⟩
10	User interruptions	⟨USER INTERRUPT⟩

The mandatory input is the name or the number of a bus stop. It is not possible to have a meaningful interaction without this information. However, all other input is regarded as optional. The second category of user inputs consists of the two ways to end the call (hang-up, explicit request). The rest of the categories include help and repeat requests, confirmations, non-mandatory advanced functionality, user interruptions and different error situations. All functionality is available with speech and dual-tone multiple-frequency (DTMF) inputs, and the system gives help on how to use these modalities.

Public pilots and laboratory tests

In order to study different aspects of the Stopman system during its development we arranged both focused laboratory tests and a public pilot lasting 30 months (Turunen *et al.* 2006). By combining the results of these studies we hoped to address the different evaluation aspects, as discussed in the previous sections. Furthermore, we wanted to study the differences between the usage in these evaluation conditions, as well as differences between the two versions of the system. Finally, we wanted to study what can be measured from automatically collected log files of calls made by anonymous users.

The initial version of the Stopman system was tested in several usability tests during the summer of 2003, and user experiences were collected during a three-month public usage test held between August and October 2003. An improved version was released to the public in November 2003. The Tampere City Transport Company had a promotion campaign on their web pages and in a newsletter that was delivered to all households in the Tampere area. In addition, an announcement about the system appeared in several local newspapers.

The first public version was in pilot use for fifteen months from November 2003 to January 2005. The number of calls to the system was fairly stable after the first three months. The exceptions were July (the holiday month in Finland), and October and March, when the system was used by the usability course participants. With this version of the system we collected 1062 dialogs (including the usability studies). The average number of calls per month was 124 for the first three months, and 52 for the rest of the months.

The second version of the system was in public use between February 2005 and 2007. In this version the names of the stops were included in addition to the stop numbers. Otherwise, this version was similar to the first version. We included to this analysis 793 dialogs from fourteen months between February 2005 and March 2006 (including the usability studies). The number of calls to the system was similar to the first version; on average there were 52 calls per month, July again being an exception.

In addition to the public pilots, Stopman was tested with the participants of the course 'Introduction to Interactive Technology' at the University of Tampere in October 2004 and October 2005. In 2004 the participants were asked to call to the system to accomplish given tasks. Each task required one call to the system. The tasks were rather simple. The main purpose was to get feedback about the use of pauses in system prompts. The calls were made to the same publicly available system. This means that most of the calls made in October 2004 were test calls.

During March 2004, the system was used as an introduction to speech applications in a study of another spoken dialog system. In the October 2005 study the participants performed slightly more complex tasks and data was collected with a copy of the system running on a separate server.

Results

In total, we collected 1855 dialogs with the system. The data collected was divided into six categories. In the first category there was the data from the first month of usage (November 2003). The second category consisted of the data from the second month (December 2003). The third category included the data from the rest of the months of the first version (January 2004–January 2005), excluding the months of the usability studies (March 2004 and October 2004). The fourth category contained the data from the months with mixed data from the real usage and the usability studies (March and October 2004). The fifth category included all data from real use of the second version (February 2005–March 2006). Finally, the sixth category contained the data from the October 2005 usability study. The categories are presented in Table 8.3.

To summarize the results, we found significant differences in the data collected during the initial use of the system, the data collected after the first two

Table 8.3 Stopman data collection categories.

	Description	Dialogs	Time
1	First month	127	11/2003
2	Second month	87	12/2003
3	First version	630	1/2004–1/2005
4	Mixed data	218	3/2004 and 10/2004
5	Second version	725	2/2005–3/2006
6	Usability study	68	10/2005

months, and the data collected in usability tests. There are highly significant differences between the first month and the rest of the pilot usage in almost all aspects of the system's use. Figure 8.1 shows a typical example.

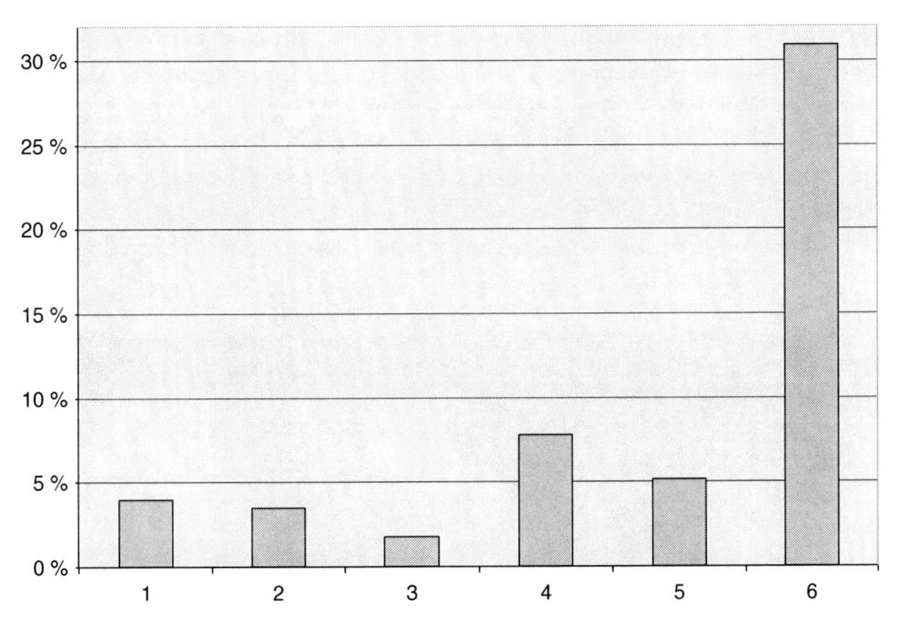

Figure 8.1 Repeat requests.

As seen in Figure 8.1, there were no significant differences in the proportion of calls with repeat requests during the real usage (cases 1, 2, 3 and 5). However, there were highly significant differences between the usability studies (cases 4 and 6) and real usage. To generalize, real users did not ask for system utterances to be repeated, while this was quite common in usability studies. Similar differences can be found almost in all the functionality categories, as presented in Table 8.2.

The only significant difference between the second month and the rest of the period of usage, however, was in the number of help requests. This suggests that usage became stable quite soon, but the users required more help during the first few months. We found no difference between the first month of the second version of the system and the rest of its usage. The differences between the system versions were quite few, while the differences between real use and usability studies were extremely high in almost all aspects.

8.3.2 TravelMan evaluation

The Stopman application was based on a server-based system. Later on, we developed a smartphone version where the speech interface was accompanied by a graphical user interface run on a mobile device (Salonen *et al.* 2005). This was further developed into the rich multimodal TravelMan application (Turunen *et al.* 2007), which provides route guidance for all types of public transport in Finland: metro, tram and bus traffic in cities and long-distance traffic in the rest of the country. In addition to providing support for planning a journey, TravelMan gives interactive guidance during the journey.

The use of TravelMan is based on multidirectional menus operated with the directional keys of the telephone. As illustrated in Figure 8.2, menus are presented using a reel metaphor: items in a menu are on top of each other and the user can roll the reel to select menu items. The currently selected node is

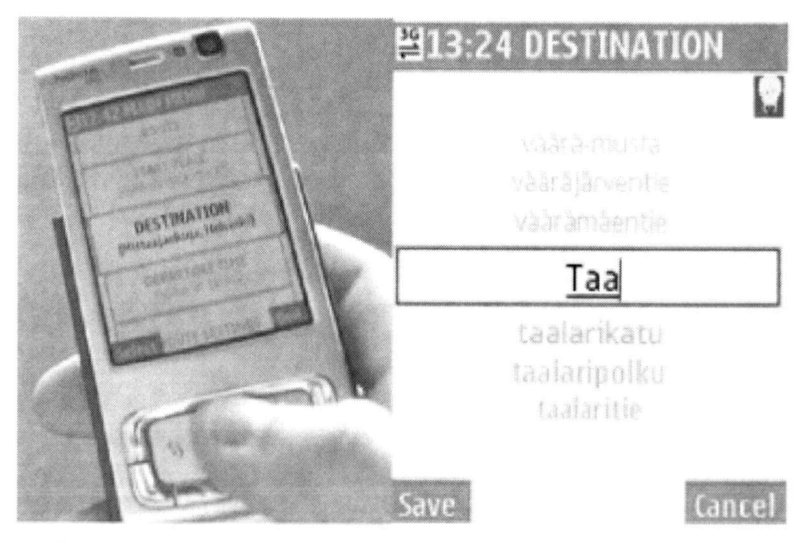

Figure 8.2 TravelMan user interface. Left: main menu with the reel interface. Right: predictive text input using the reel interface.

enlarged to provide more information, making it easier to see the information on the small display.

The interface design is inspired by focus-and-context techniques, such as Fisheye Menus 7. The reel-interface, however, is two-dimensional, uses a reel instead of menus, and is tightly integrated with speech outputs. The content of each item in a reel is read out loud by the speech synthesizer when the item is activated. However, the spoken content is not necessarily the same as the content presented on the display, since speech and text have different strengths and weaknesses. Speech outputs use full sentences to keep the message easily comprehensible. The temporal nature of speech also makes it possible to skip the ends of spoken prompts by quickly moving to the next item. Because of this, presenting the most important information first can speed up the usage.

In the journey-planning phase, a user enters the departure and destination addresses or locations (street names and numbers, places of interest and city names) using one of the available input methods: speech input and two variations of text input. Speech input is based on using the keyboard to start and stop the recoding of audio. For speech recognition, the application uses a distributed system, where the audio is sent over the network to a server-based Finnish language recognizer. The language model for speech recognition consisted of 8896 street names and addresses, 2053 place names, and numbers from 1 to 100, and totaled 11049 words.

The second way to enter addresses is two variations of text input. In addition to normal multi-tap text entry, predictive text input optimized for the address data entry is available. The language model for the prediction was the same set of street and place names as used in the speech recognition system. The reel interface is used in predictive entry mode so that initially the reel contains all valid addresses, but updates in when the user enters new characters. At any time, the user can select a street name from the reel. When a street name has been selected or completely entered, the predictive text input automatically switches to number mode for entery of the street number. The text input is designed to be fully accessible without seeing the screen. When the user types characters the system speaks out loud the most likely character sequences according to the current domain. When there is only a reasonable number of results left, the system informs the user with sound and vibration. This is performed in the background so the user can continue typing, or select the correct address using the reel interface.

In applications such as TravelMan the entering of information is challenging because of the limited keypads and the difficulties of the mobile usage context, as discussed in the previous sections. Here we wanted to study how efficient speech can be compared to regular text input and highly optimized predictive, domain-specific text input. This was motivated by previous research that shows that that there are several speech-based input methods that can be faster than generic predictive text-input and regular multi-tap text input. Still, little is known on their true efficiency. Furthermore, there are few studies on user acceptance of different input methods, and speech in particular is often dismissed for very small vocabulary cases. Finally, there has been always the question of acceptance of synthesized speech output. In mobile applications, speech output is considered particularly useful for visually-impaired users. However, it can be helpful to all users, since in the mobile context of use we all sometimes have limited vision, for example due to a need to monitor our surroundings. There are speech synthesizers in some new mobile phones, but little is known of their usefulness, usability and user acceptance. Here, we wanted to study the objective and subjective factors affecting these considerations.

Evaluation of user expectations and experiences

We used a variation of a subjective evaluation method called SERVQUAL (Parasuraman *et al.* 1988) in user studies of the TravelMan application. The SERVQUAL method was originally developed for evaluation of traditional services and was later used in the evaluation of a spoken-dialog system. More recently, it was modified to be used in speech-based (Hartikainen *et al.* 2004) and multimodal applications. In this section we show how the basic principles of the SERVQUAL method can be used to study user expectations and experiences of multimodal mobile speech applications.

The basic idea is to collect user expectations before the use of an application, and compare the results with user experiences collected after use. This is done with a questionnaire containing various statements related to the quality of the application. For example, one statement can be 'Predictive text input is quick to use'. For each statement, users mark two values between 1 and 7, an acceptable level and a desired level of quality. After use, the users record their actual perceived level. In this way, the method produces a gap between pre-test user expectations the post-test user experiences. Figure 8.3 illustrates these concepts. For the example user, the accepted level for multi-tap entry is 3, the desired level is 7, and the perceived level is 2.

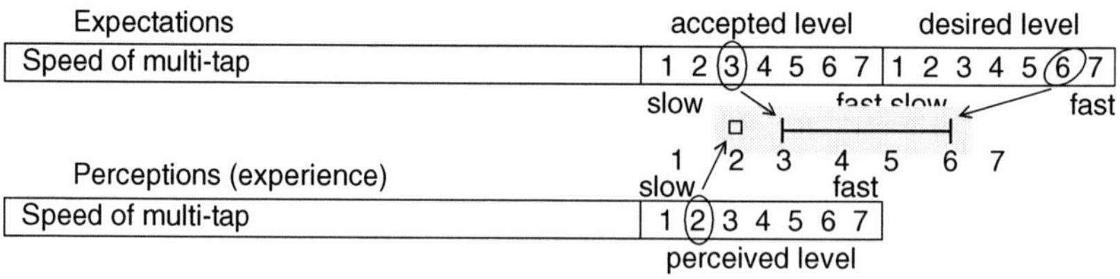

Expectations
accepted level
desired level
Speed of multi-tap
1 2 3 4 5 6 7
1 2 3 4 5 6 7
slow
fast
1 2 3 4 5 6 7
slow
fast
Perceptions (experience)
Speed of multi-tap
1 2 3 4 5 6 7
perceived level

Figure 8.3 User expectations and perceptions.

The gap can be expressed using two disconfirmation measures, the measure of service superiority (MSS) and the measure of service adequacy (MSA). MSS measures the difference between the perceived level and the desired level, and MSA the difference between the perceived level and the accepted level. If experiences are in the range of expectations, MSS values are negative and MSA values are positive. The range spanning the accepted and the desired levels is called the zone of tolerance. For the example in Figure 8.3, The zone of tolerance is [3, 7], MSS is −4, and MSA is −1, meaning the perceived user experience is not within the zone of tolerance and is below the accepted level.

The SERVQUAL method is particularly suitable for iterative development and prototyping, since it has been designed to provide information that can be used when further developing services. Most importantly, it indicates what the strong points of the application are, and where further effort is needed. For example, the results could imply that it is more critical to concentrate on intuitiveness than on performance.

Test setup

The TravelMan evaluation contained 38 students (27 male, 11 female) from the local university. Their ages ranged from 18 to 45 years. Twenty-one participants had prior experience with Series 60 smartphones, which was the type used in the evaluation. Objective and subjective metrics were collected in order to analyze the interactions and elicit feedback from the participants.

Participants were introduced to the TravelMan application by a web-based wizard. The main features of the TravelMan application were described, but the actual usage instructions were not revealed at that point. After introduction, pre-test user expectations for acceptable and desired levels of quality were gathered using a web- form questionnaire based on the SERVQUAL method.

The test took place within from one to two weeks of the introduction and recording of the subjective questionnaires. A Nokia N95 smartphone (which has a numeric keypad) was used in the test, and data was gathered with the test application's internal logging system – this stored every key press and action. The participants were given 4 exercise tasks and 21 evaluation tasks, 7 for each modality. The exercise tasks did not include the use of any input method for addresses (speech, multi-tap text, predictive text); there was only general navigation to familiarize the participants with the reel user interface. The actual tasks were designed in a way that participants had to enter either the start location or the destination in each task, but the actual answer (for example, the time the next bus departed) was not to be found in the address input view. If an initial attempt to accomplish a task failed (e.g. due to speech recognition error or a mistyped address), the participant was asked to try again.

At this point, the participants were not informed that the test was related to input methods, but instead were told that it was a regular usability test to discover problems in the software.

The evaluation was organized as a within-subject study. The three tasks sets were the same for all participants and the order of modality presentation was randomized so as not to introduce any learning biass. The pairing of task set to modality depended on the group. The tasks were always presented in the same order for the different modalities, and the addresses in each task were predetermined, in order to keep the tasks comparable. In this way, each modality was used equally during the whole test. Task data entry methods were not used in the experiment. After completing the task set with the given modality, the participants were asked to fill in a questionnaire. The questions were exactly the same as in the pre-test questionnaire, but this time the participants gave only one value, based on their experience of use.

Since TravelMan uses server-based speech recognition, speech input response times are heavily affected by the connection speed between the mobile device and the server. Twenty-four participants used 802.11g WLAN connections, nine used 3G connections (max. 384 kbit/s, but in practice a lot of variation), and five used EGPRS connections. This mostly affected the speech tasks, since transferring the audio data to the server took more time than transferring text input strings. For speech, total times for task completion varied from 5–237 sec. Input time and transfer time are reported separately, where appropriate, and considered in the analysis of the results.

Objective and subjective results

We collected several objective usage metrics, focusing on task-completion times. As shown in Table 8.4, the median task completion time was 13.7 sec for speech input, 17.2 sec for predictive text input, and 30 sec for multi-tap text input. Speech recognition rates varied greatly between different users. The range was from 100% to about 45%. The overall recognition rate was 70.4% and 97% of recognition task cases were completed successfully within three attempts.

Table 8.4 Average input entry times per modality.

Modality	Number of tries	Input time	Transfer time	Total time
Multi-tap	1.02	28.8	0.8	30.0
Predictive	1.05	15.6	1.1	17.2
Speech	1.47	5.7	7.8	13.7

As seen from Table 8.4, speech is the fastest input method, even with the long transfer times caused by distribution and error-prone speech recognition. Theoretically, it could be even better: the time for giving a single entry with speech was 4.6 sec (SD 2.2), the mean for audio recording time being 3.3 sec (SD 1.5). For normal and predictive text input, the corresponding times were 28.2 sec (SD 16.9) and 14.8 sec (SD. 31.5), respectively. Next, we were interested to find out how this correlated with subjective metrics.

We calculated MSS and MSA measures, as discussed previously, for different usability dimensions (speed, pleasantness, clarity, error-free use, error-free functionality, learning curve, naturalness, usefulness, and willingness to use in future) for each modality (multi-tap text input, predictive text input, speech input, speech output) and the application itself.

When comparing the main dimensions–usefulness and future use–there were significant differences between modalities, as demonstrated in Table 8.5, There are cases, such as the perceived usefulness of TravelMan and predictive text input, where MSS is positive or very near to zero. In these cases, the perceived values are very high (above 6) in an absolute terms as well. Furthermore, MSA was relatively high in these cases (more than 2.5), indicating significant differences in user expectations and experiences. To conclude, users found predictive text input and TravelMan to be very useful, and better than their expectations.

Table 8.5 Subjective usefulness measure.

	Accepted	**Perceived**	**Desired**	**MSA**	**MSS**
Multi-tap	3.42	5.13	5.84	1,71	-0.71
Predictive	3.68	6.26	6.32	2.58	-0.10
ASR	2.76	5.05	5.55	2.28	-0.50
TTS	3.11	4.00	5.71	0.89	-1.71
TravelMan	3.42	6.08	6.05	2.66	+0.03

In other cases, MSS and MSA values were as expected, below and above zero, respectively. In the case of multi-tap text input, users found it quite positive. In the case of speech input, experiences were quite close to desired levels, and users found it quite useful, but since expectations were not too high concerning future use, it was less favored than text inputs. Regarding speech output, users expected it to be more useful than speech recognition, but perceived its quality to be less, both in relative and absolute terms. Furthermore, MSS was quite high for speech outputs when future use is considered, even with mediocre expectations.

When different usability factors and the willingness to use the TravelMan application and its different modalities are considered, there were highly significant correlations between the perceived usefulness and the willingness to use the same modality and application, but no others. In other words, if users found TravelMan or any modality to be useful, they were also willing to use it. In general, no single usability factor correlates with willingness to use. For usefulness, there is a clear explanation: perceived naturalness has a strong correlation with usefulness. Furthermore, other usability factors correlate with naturalness equally. In other words, the relationship between subjective ratings can be defined using the user experience model illustrated in Figure 8.4.

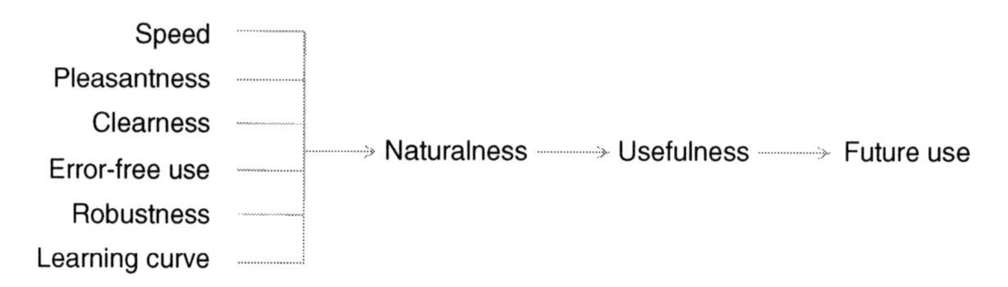

Figure 8.4 User experience model.

In our study, the user experience model was true in all but three cases, which makes these particularly interesting. One exception was discovered in the correlation between the pleasantness of speech output and expectations of future use. Based on this, it seems that the lack of pleasantness is a deciding factor in users' assessment of speech output. In addition, two positive correlations were found when different modalities and the Mailman application were compared. These were between the willingness to use speech input and output in future, and between the willingness to use speech output and the TravelMan application. Based on these results, a willingness to use speech input and output seems to be related, and a willingness to use synthesized output affects the willingness to use the TravelMan application. The latter finding is quite important, and suggests that speech synthesis should be used with care in similar applications.

Subjective and objective metrics

One of our aims was to find correlations between objective and subjective metrics in a somewhat similar way to some of the evaluations performed with the PARADISE method. However, we were able to find only few direct correlations. For example, we found only a weak correlation between the actual speech recognition accuracy and the perceived accuracy of speech inputs, and no correlation between speech input accuracy and the perceived usefulness of speech input. Even users with 100% speech recognition accuracy used the full scale in their ratings. Furthermore, objective metrics were not able to explain the perceived usefulness or willingness to use different modalities, the only exception being a negative correlation between the total time taken for predictive text input in the last task and the perceived usefulness of speech input. These findings suggest that in multimodal applications, where the use of the application is not dependent on any single modality, it is hard to correlations between objective metrics and user satisfaction similar to those seen in unimodal applications, where the single modality dictates the usefulness of the application.

8.3.3 Discussion

Both evaluation examples presented show the effect of mobile use context, or lack of it, on results. In the case of the Stopman system, there were significant differences between the laboratory studies and the field study in almost all aspects examined. Naturally, there are many other factors that affect the results. For example, in field studies the users learn to use the system more effectively, and after a while irrelevant calls disappear, as noted in other studies. In usability studies, the tasks may not have real meaning for the users, and the users are sometimes even too co-operative. Interestingly, a very similar study was conducted for the Let's Go System! (Ai *et al.* 2007). Is is particularly noteworthy that these results support our findings in many cases, but there are contrary results as well. This suggests that the nature of the system and the interaction affects the results, even with very similar systems as in these cases.

The TravelMan study, although performed in a laboratory, resembled the situation where a user is planning a route at home before starting a journey, and so in this case it was not real mobile study. However, it failed to provide any strong motivation for speech input and output. In mobile situations, the hands- and eyes-free interaction may favor speech and limit the usefulness of other methods. For example, pen-based soft-key text input has been shown to be slower while walking. Furthermore, in our laboratory conditions, speech and pen input did not provided any added value. Overall, it will be a major challenge to assess the performance and user experience of multimodal speech applications in situations where a user is on the move in varying social and physical conditions, as discussed in previous sections.

Overall, these experiments show that laboratory studies can be used to measure efficiency, but the results should be considered carefully when used to model interactions taking place in mobile environments. Furthermore, as the TravelMan study shows, the relationship between efficiency and user satisfaction is quite complex, and it is unlikely that these attributes correlate with one another universally, as also discussed in the general usability literature (Bernsen and Dybkjær 2000). For example, faster and less error-prone methods are not always the preferred ones, as shown in studies of spoken-dialog systems where users prefer system-initiative dialogs over more efficient user-initiative dialogs (Walker *et al.* 1998). Some researchers even claim that user perceptions do not correlate with objective measures and our study with the TravelMan system supports this claim.

8.4 Theoretical measures for dialog call-flows

Algorithm design and analysis rapidly attained maturity with the introduction of time and space complexity measures. Dialog call-flow design and analysis for pervasive devices is a relatively nascent art, complicated by the variation in resource constraints. We investigate the problem of dialog call-flow characterization for pervasive devices, with the objective of defining complexity measures for dialog call-flows.

A dialog call-flow executing on a pervasive device is an interplay among the *device*, the *human* and the *speech system*, and can therefore be completely characterized by the $\langle \underline{r}esource, \underline{u}sability, \underline{t}echnology \rangle$ triple. Typical examples of *resource* are memory and energy; *usability* is indicated in the number of questions; and *technology* is exemplified in the accuracy of the speech recognition system. These are charateristics of the *call-flow*, but the call-flow *characterization* itself is {device,human,speech system}-*independent*.

We instantiate $\langle r, u, t \rangle$ with$\langle \underline{m}emory, \underline{q}uestions, \underline{a}ccuracy \rangle$ to introduce the $\langle m, q, a \rangle$-*characterization* of dialog call-flows. We select m, q, a metrics to define various $\langle m, q, a \rangle$-*complexity* measures for a call-flow. Every call-flow thus has a complexity measure associated with it – a feature indispensable for the systematic analysis and design of dialog call-flows.

8.4.1 Introduction

The proliferation of pervasive devices has stimulated the development of applications that support ubiquitous access via multiple modalities. Speech as a medium is especially convenient for users on the move. Therefore, more and more applications are becoming speech-enabled. Speech-enabling an application requires designing a dialog call-flow. Dialog call-flow design is an art. One of the primary reasons for this is the inherent challenge of designing human interfaces. The intricacies of usability, which has many subjective elements, have to be carefully considered while designing a call-flow. Designing dialog call-flows for pervasive devices is further complicated by the resource restrictions imposed by pervasive devices, and the increasing capabilities of such devices (Rajput *et al.* 2005). From this pervasive point of view, what essentially characterizes a call-flow? The isolation of the characteristics that distinguish one call-flow from another would greatly facilitate the analysis and design of 'pervasive' call-flows.

An algorithm is a computer-independent abstraction of a computer program. An algorithm is characterized by the memory it uses and the time (number of steps) it takes to execute. Since the actual memory and actual time it takes depends upon the implementation (program) and the machine on which it runs, the machine-independent characterization of memory and time give the space-complexity and the time-complexity of the algorithm. These complexity measures are useful for comparing the complexity of two algorithms, and have systematized algorithm analysis and design (Knuth 1998).

The goal of this section is to identify analogous parameters and define complexity measures for dialog call-flows. Unlike algorithms, a dialog call-flow is an interface for interacting with humans. Therefore, subjective and hard-to-quantify factors–whether the call-flow is intuitive, the conversation fluent, the voice pleasing–become important. However, our primary interest is in identifying quantifiable characteristics of call-flows designed for interactions on pervasive devices.

The advantage of this approach is two-fold: the identification of a *characterization* of dialog call-flows and the definition of various *complexity* measures based on this characterization. A characterization defines a template that is instantiated filled. Since we want to characterize call-flows, the characterization itself should be device (human, system)-independent. In the case of time and space characterizations of algorithms, an abstract machine is considered so that the characterization is not specific to any implementation or machine. Our characterization is fundamental, but not exhaustive. It is fundamental because it captures the primary design trade-offs–fewer questions imply larger grammars and require more memory–but not exhaustive because it does not include the subjective-but-important factors mentioned above. This also suggests a two-stage approach to call-flow design. Once the primary parameters have been determined, the second stage can involve decisions on the 'soft' subjective considerations. The dialog call-flow complexity measures provide a quantifiable basis for comparison between two dialog call-flows. It is possible to define various complexity measures based on a characterization. The selection of the appropriate complexity measure depends upon the application requirements. The fact that the characterization is be device-independent also implies that the units of complexity measurement should be abstract.

Many approaches have been suggested for evaluating call-flows. Since a complete evaluation must include the subjective factors mentioned above; one cannot expect a consensus on these evaluations. The evaluations are likely to depend on the particular application being designed. PARADISE (Walker *et al.* 1997) is a well-accepted evaluation framework that has been used to compare dialog systems (Walker *et al.* 1998). It provides a framework whose parameters can be configured to meet the specific evaluation requirements. Danieli and Gerbino (1995) have suggested metrics that should be used to evaluate and compare dialog strategies. However, these evaluation metrics and the framework were applied on a deployed system with the aim of capturing the effectiveness with which users responded to the system. Although this is the final goal of any voice application, we suggest introducing characterization to aid the call-flow design process before it is actually implemented and deployed. A key feature of our idea is the identification of quantifiable, fundamental design considerations. In the design lifecycle, this can be followed by soft design, and then evaluated by any measure that is deemed appropriate for the application.

The next section describes the details of the characterization and its complexity. The following section illustrates the effectiveness of a characterization-based approach for evaluation and shows how to compute the complexity of a dialog call-flow through an example.

8.4.2 Dialog call-flow characterization

We first describe the features of a call-flow. Then we introduce the $\langle m, q, a\rangle$-characterization of a call-flow, and show how the characterization parameters are derived from the features. Based on the characterization, various $\langle m, q, a\rangle$-complexity measures can be defined. We conclude this section with a sketch of where this analysis fits in the voice application lifecycle.

Call-flow features

A dialog call-flow consists of a sequence of questions and expected answers. The questions are represented through prompts in the VXML. The expected answers are encapsulated in a grammar. Each prompt has a grammar associated with it, unless the prompt is only providing some information to the user, without expecting any input. The structure of the call-flow decides the order in which the questions are going to be asked. A call-flow structure can be sequential, so that the next question to be asked is always fixed. On the other hand, for a tree-structured call-flow, the next question to be asked will depend on the user input to the previous question. The *prompts, grammars* and the *structure* are the features of a call-flow. The characterization parameters will be extracted from one or more of these call-flow features.

8.4.3 $\langle m, q, a \rangle$-characterization

A dialog call-flow executed on a pervasive device is an interplay among the *device*, the *human* and the *speech system*, and can therefore be completely characterized by the $\langle \underline{r}esource, \underline{u}sability, \underline{t}echnology \rangle$ triple. For a pervasive device, the memory size and energy are the main *resource* constraints. From the *usability* perspective, two objective considerations are the number of prompts and the time it takes to execute the call-flow. A speech recognition system (*technology*) is largely characterized by its accuracy, which depends upon multiple factors including the size of its vocabulary, the training set, and the particular grammar in question. The energy consumed by a dialog call-flow is likely to be a complex function of the memory and the duration of a call-flow. The number of questions is roughly an indicator of the time it takes for the call-flow to execute. We consider $\langle m, q, a \rangle$ to be fundamental for two reasons: (1) their direct correspondence with call-flow features (taken together they cover all the call-flow features), and (2) because other parameters (such as energy required for the call-flow) may be complex, albeit unknown, functions of these. Informally, these three independent factors define a three-dimensional space where other factors can be expressed as functions of them.

- Memory m. We consider the measurement of maximum memory that is required by the call-flow at a particular point of execution. For this, we identify the various call-flow features that lead to changes in memory requirement within the call-flow. The only call-flow feature which causes a change in the memory requirement is the grammar. The number of choices in a grammar has a direct correlation with the memory that is required to process the utterance. A speech recognition system matches the user utterance with the choices provided by the grammar and so if there are more choices in the grammar, the recognition system requires more memory to generate confidence scores for all those choices. We define the maximum memory m of a call-flow as the maximum number of choices in a particular grammar of the call-flow.

$$m = \max_i (c_i), \tag{8.1}$$

where, c_i is the number of choices in grammar G_i. This number is specific to a particular grammar within the call-flow. It limits the execution of the call-flow to devices that can support the memory required by this grammar. Moreover, by measuring the memory in terms of a number of choices, the characterization parameter m is independent of the lower-level speech recognition implementation and device details.

- Questions q. We represent the number of questions that a call-flow has to execute as q. The q value of a call-flow thus measures the *order* of time that is required for the call-flow to execute. The number of questions that are asked in order to perform a task defines the *length* of the call-flow. A sequential call-flow has a fixed length for q. For a tree-structured call-flow, since the number of questions asked depends upon the answers given by the user, various measures, such as the maximum value, average value, median value, of the length of a call-flow may be meaningful. Therefore, the following are sample ways (there may be other statistical measures, for example) to measure the characterization parameter q.

$$q = N_q$$

$$q = \max_i N_i$$

$$q = \frac{\sum_i^n N_i}{n}$$

  where N_q is the number of questions in a sequential call-flow and N_i is the number of questions in the ith path for a tree-type call-flow.

- Accuracy a. The *perplexity* of a grammar is a representation of the speech recognition accuracy at the text level: the higher the perplexity, the less the accuracy. For a given call-flow, the grammar with the maximum perplexity provides the lower bound for speech recognition accuracy of the call-flow. This provides a measure of the characterization parameter a. Alternatively, for a given call-flow, the total number of unique words contained in all the grammars represents the size of the *acoustic space* of the call-flow. Clearly, the larger the acoustic space, the harder the recognition task. Perplexity captures the recognition hardness for a particular grammar in the call-flow and is therefore a local measure, while the acoustic space size provides a global measure. These measures provide a quantification of the characterization parameter a.

$$a = \max_i P_i$$

$$a = |\bigcup_i w_i|$$

  where, P_i is the perplexity for grammar i and, w_i is the set of unique words in grammar i.

Table 8.6 summarizes the relationship between the characterization parameters and call-flow features. It shows that while memory is dependent only on

Table 8.6 Characterization parameters and their dependency on call-flow features.

Call-flow features	**Characterization parameters**		
	Memory m	Questions q	Accuracy a
Prompt		*	
Grammar	*		*
Structure		*	*

the grammar, accuracy is a function of grammars as well as the of the call-flow. structure of the call-flow.

8.4.4 $\langle m,q,a \rangle$-complexity

The $\langle m, q, a\rangle$-characterization provides a basis to define various $\langle m, q, a\rangle$-*complexity* measures. An $\langle m, q, a\rangle$-*complexity* measure is a group of quantities that can be associated with a call-flow. While the characterization identifies the parameters, the complexity quantifies the numbers associated with a particular call-flow. Since multiple quantities may be required for expressing a single parameter, there is no generally unique $\langle m, q, a\rangle$-*complexity*. For example, a tree-structured call-flow has an average length as well as a maximum length. Depending upon the particular application, any subset of these might be required to adequately capture the complexity of call-flows. Two sample $\langle m, q, a\rangle$-*complexity* measures for a tree-structured call-flow are: $\langle m, N_q, \langle \max_i P_i, |\bigcup_i w_i| \rangle\rangle$-*complexity* and $\langle m, \langle \max_i N_i, \frac{\sum_i^n N_i}{n}\rangle, \max_i P_i\rangle$-*complexity*. The former has two numbers to capture accuracy, while the latter has two to capture the questions.

Role of characterization in the application lifecycle

Figure 8.5 shows the typical lifecycle of a voice application. The application is defined through a set of voice user interface (VUI) specifications. The design of the application uses the VUI specifications and then the design is implemented. The feedback of the testing phase is passed on to the implementation team and there are a few iterations. Typically, once the testing is over, the voice application is deployed. It is only after deployment that the various evaluation experiments are performed and user feedback is obtained. The evaluation provides a mechanism for comparing different call-flows and the feedback can be used to improve the design and implementation. In the case of call-flow design

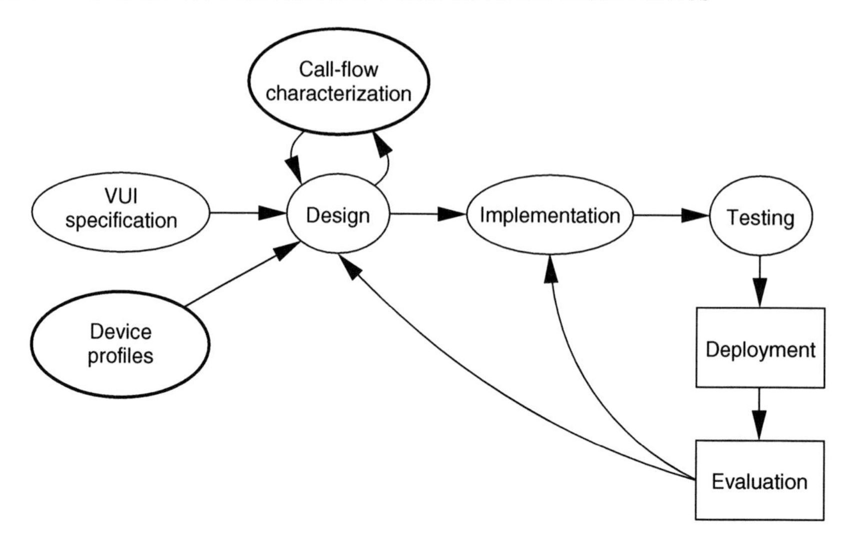

Figure 8.5 The voice application lifecycle.

for pervasive devices there are two additional components. First, there are the *device profiles*, which consist of the capabilities of devices including details such as screen size, battery life, memory, GPRS-enablement, etc. From the set of device profiles we can extract a 'lowest common denominator', for example the lowest memory among all the devices, and use this to drive the call-flow design. This requires a proper parameterization of the $\langle m, q, a \rangle$-complexity of the call-flow being produced. The second additional component, call-flow characterization/complexity is intrinsic to the design phase of the lifecycle.

8.4.5 Call-flow analysis using $\langle m, q, a \rangle$-complexity

A call-flow design follows the VUI specifications that are usually provided by VUI experts after taking into consideration the functionality of the voice application. The design is completely driven by the VUI specifications, and once the design is complete it is implemented using one of the voice markup languages such as VoiceXML or SALT. The implementation can be performed using a J2EE/JSP architecture if the application involves generation of many dynamic voice pages–similar to the way in which dynamic HTML is generated for web applications. The voice application still runs on a voice browser that interprets the VoiceXML language, but the dynamic java servlet pages (jsps) are handled at the HTTP browser. However the call-flow design is independent of the implementation and deployment environment. Thus, a characterization of call-flows at the design level provides a mechanism to compare call-flows at a much higher level of abstraction, without going into details of its implementation. Here we present two application call-flows to demonstrate the inferences that can be drawn by analysing their $\langle m, q, a \rangle$-complexity.

Figures 8.6 and 8.7 show the call-flows of an airline enquiry system. The application requires four values from a user: date of travel, departure airport, arrival airport and the preferred airline. Based on the input received, the application forms a query and fetches relevant results from the back end. Though both the call-flows have the same functionality, they have been designed separately so that they have different memory requirements and different usability features. While the first call-flow is more concise in its conversation, the call-flow in Figure 8.7 has a smaller memory requirement since it uses smaller grammars. The numbers in parentheses in Figures 8.6 and 8.7 represent the memory (in terms of number of choices) required by the grammar for that particular question. We now show how to compute the m, q, a values for these two call-flows.

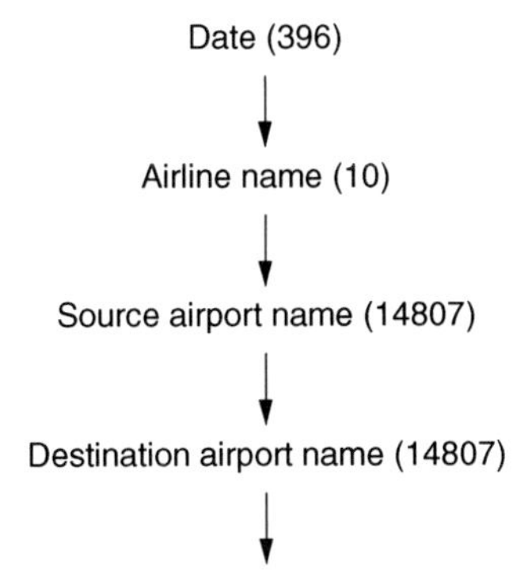

Figure 8.6 A sample airline enquiry call-flow design.

Four different grammars are used in Figure 8.6. The characterization parameter m is the maximum number of choices in a particular grammar within the call-flow, in other words $\max(24, 10, 14\,807, 14\,807) = 14\,807$. The parameter q in case of this sequential call-flow is 4, the number of questions in the call-flow. To measure the perplexity component a, we have to measure the perplexity of each grammar, the grammar with highest perplexity being reflect in this component. The second component, acoustic space size, can be measured by counting the total number of unique words that exist in all the grammars in the call-flow. The perplexities for the four grammars used in Figure 8.6 are (24, 10, 14 807, 14 807), assuming that there is no language syntax associated with the airline and airport grammars and that the date grammar specifies the structure of specifying a date. The total number of unique words combining all the four grammars are 14 912. Thus, a can be encapsulated as $\langle 14\,807, 14\,912 \rangle$.

Note that an ideal call-flow should have lower numbers for each of the three parameters, irrespective of the way they are measured, a minimum memory requirement m, minimum number of questions q in the call-flow and a very high accuracy, in other words less perplexity or less number of words in the call-flow, a. The $\langle m, N_q, \langle \max_i P_i, |\bigcup_i w_i| \rangle\rangle$-*complexity* of the call-flow is $\langle 14\,807, 4, \langle 14\,807, 14\,912 \rangle\rangle$.

On the other hand, the call-flow in Figure 8.7 has a $\langle m, q, a \rangle$-complexity characterization of $\langle 396, 6, \langle 50, 15\,763 \rangle\rangle$ (assuming there is a total of 50 states and 500 cities). The call-flow does not use a airport grammar since there is a large number of airports and this increases the memory requirement of the call-flow. So it first asks the user for the state in which the airport exists. Knowing the state, it asks for a city name within this state and so it uses a smaller state-specific grammar. Once the city has been provided, the system uses a city-specific grammar that has the names of airports only in that city. This reduces the number of choices in the eventual airport grammar. A similar technique can be used while asking for the destination airport too, but this has not been shown in the figure. This call-flow asks a maximum of six questions. As seen in Figure 8.7, some of the questions may not be required–the ones

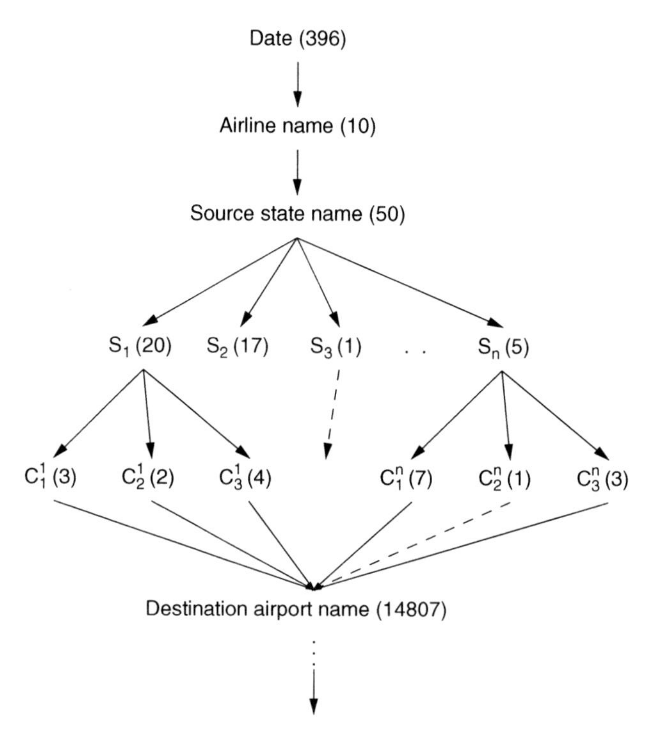

Figure 8.7 Tree-structured call-flow design for airline enquiry.

shown by dashed lines in the figure. If there is only one city in a particular state the user will not be asked to specify the city within the state. Similarly, if there is only one airport within a city, then the user will not be asked for the airport name. Depending on the user's answer, the call-flow may require only four questions for completion. Thus, an average number of questions can also be used as a measure for q for this tree-structured call-flow.

From the complexities of the two call-flows, it is clear that while the first call-flow requires fewer questions, it requires more memory, the opposite of that in the second call-flow. Moreover, although the second call-flow has a better accuracy (perplexity), it requires a higher number of questions and a larger acoustic space to complete the dialog. These two call-flows illustrate the fundamental trade-offs in dialog call-flow design.

8.5 Conclusions

Mobile devices have become common and new applications and devices become available constantly. Mobile use results in new usage contexts and new factors that must be considered in the design and evaluation of such systems. However, it is not trivial to determine and analyze these factors.

We can speculate on various factors that the mobile context provides and discuss how to model these in laboratory conditions, when sufficient. The actual variables are found in real life field tests, and some of these can be studied in more controlled experiments in laboratory studies. However, new mobile devices and services change the usage context all the time, so we are aiming at a moving target. Furthermore, it is becoming more important to focus on overall evaluation of applications rather than single factors affecting the usability and efficiency. As shown in the example studies in Section 8.3, the overall user experience is very hard to measure from individual factors, since everything affects the overall quality of the system. Furthermore, individual differences and preferences are becoming more important with new applications, such as those targeted towards entertainment and other leisure-related activities.

Some variables that can be deduced and in some cases can be found in the existing research literature include:

- physical movement;
- tiredness (walking in a lab, treadmill etc.);
- mental load (cognition);
- noise (we can add background noise to a laboratory);
- lighting conditions (we can simulate this to some extent);

- temperature (again, we can simulate this to some extent);
- movement of the environment (subway train etc.);
- social context (very hard to model, scenario-based testing procedure may provide some of the effects; this is a very complex area);
- etc.

In the field we can undertake qualitative studies where we can find new and important variables that would not be spotted in laboratory. We can later try to simulate these in the laboratory if we need controlled experiments. Usability evaluations can take place in the field, since they are qualitative in nature. As in usability evaluations, field tests are very important; we can speculate on the factors of mobile context but we can only see what factors there really are when we go into the field. Field studies are very challenging and demanding but invaluable. One must also remember that field studies should be conducted throughout the product cycle, starting from the requirements gathering phase and possibly ranging to log studies after deployment to provide data for further iterations.

We introduced the $\langle m, q, a\rangle$-characterization of dialog call-flows for pervasive devices. Based on this characterization, it is possible to define a family of $\langle m, q, a\rangle$*-complexity* measures. Once a suitable complexity measure has been defined for an application, we have a basis to compare two dialog call-flows. This is a crucial step in dialog call-flow design. The measure forms the basis of call-flow adaptation for pervasive devices. We believe that our approach is a first step towards the systematization of the analysis and design of dialog call-flows, and hopefully will play a role akin to assessment of time and space complexity in the field of algorithms. It is important to reiterate that this characterization, although fundamental, is not complete. A complete characterization requires the consideration of subjective factors that are beyond the scope of this chapter. We consider this characterization fundamental because it takes into account the very basic elements of a call-flow and illustrates the trade-offs involved.

Bibliography

Ai, H., Raux, A., Bohus, D., Eskenazi, M. and Litman, D. (2007). Comparing spoken dialog corpora collected with recruited subjects versus real users. In *Proceedings of SigDial 2007*.

Anderson C. A. and Bushman B. J. (1997). External validity of 'trivial' experiments: The case of laboratory aggression, *Rev. Gen. Psychol.*, 1, 19–41.

Baillie L and Schatz R (2005) *Exploring Multimodality in the Laboratory and the Field*. In *ICMI 2005*.

Barnarda, L., Yia, J.-S., Jackoa, J. A. and Sears, A. (2005) An empirical comparison of use-in-motion evaluation scenarios for mobile computing devices. *Int. J. Human-Comp. Stud.*, 62, 487–520.

Bernsen, N. O. and Dybkjær, L. (2000) A methodology for evaluating spoken language dialogue systems and their components. In *Proceedings of the 2nd International Conference on Language Resources and Evaluation, LREC 2000* (Athens, Greece), 183–188.

Betiol, A. H. and de Abreu Cybis, W. (2003) Usability testing of mobile devices: a comparison of three approaches. In *INTERACT 2003, LNCS 2585*, 470–481.

Bolger, N., Davis, A. and Rafaeli, E. (2003). DIARYMETHODS: capturing life as it is lived. *Annu. Rev. Psychol.* 54, 579–616.

Bracht G. H. and Glass, G. V. The external validity of experiments. *Am. Educ. Res. J.*, 5, 437–474.

Brandt, J., Weiss, N. and Klemmer, S. R. (2007) txt 4l8r: Lowering the burden for diary studies under mobile conditions. In *CHI 2007*.

Brewster, S. (2002) Overcoming the lack of screen space on mobile computers. *Pers. Ubiquitous Comp.*, 6, 188–205.

Brewster, S., Chohan, F. and Brown, L. (2007) Tactile feedback for mobile interactions. In *CHI 2007*.

Carlsson, V. and Schiele, B. Towards systematic research of multimodal interfaces for non-desktop work scenarios. In *CHI 2007*.

Danieli, M. and Gerbino, E. (1995) Metrics for evaluating dialogue strategies in a spoken language system. In *Proceedings of the AAAI Spring Symposium on Empirical Methods in Discourse Interpretation and Generation*, 34–39.

Duh, H. B.-L., Tan, G. C. B. and Chen, V. H.-H. (2006) Usability evaluation for mobile devices: a comparison of laboratory and field tests. In *ACM MobileHCI 2006*, Helsinki, Finland, 181–186.

Fithian, R., Iachello, G., Moghazy, J., Pousman, Z. and Stasko, J. (2003) The design and evaluation of a mobile location-aware handheld event planner. In *MobileHCI 2003*, LNCS 2795, 145–160.

Gorlenko., L. and Merrick. R (2003) No wires attached: usability challenges in the connected world. *IBM Syst. J.*, 42(4), 639–651.

Hartikainen, M., Salonen, E.-P. and Turunen, M. (2004) Subjective evaluation of spoken dialogue systems using SERVQUAL method. In *Proceedings of the 8th International Conference on Spoken Language Processing, ICSLP 2004,* Jeju, Korea. ISCA, 2273–2276.

Hirschman, L. and Thompson, H. S. (1996) Overview of evaluation in speech and natural language processing. In R. A. Cole, J. Mariani, H. Uszkoreit, A. Zaenen and V. Zue (eds), In *Survey of the State of the Art in Human Language Technology*, Cambridge University Press (March 13, 1998).

Holmquist, L. E., Höök, C. Persson, P. (2002) Challenges and opportunities for the design and evaluation of mobile applications. Presented at the workshop: In *Main Issues in Designing Interactive Mobile Services*, ACM Mobile HCI.

Hone, K. S. and Graham, R. (2000) Towards a tool for the subjective assessment of speech system interfaces (SASSI). *JNLE*, 6, 287–303.

Hone, K. S. and Graham, R. (2001) Subjective assessment of speech-system interface usability. In *Proceedings of the 7th European Conference on Speech Communication and Technology, Eurospeech 2001,* Aalborg, Denmark.

Isomursu, M., Táhti, M., Váinámó, S. and Kuutti, K. (2007) Experimental evaluation of five methods for collecting emotions in field setting with mobile applications. *Int. J. Human-Comp. Stud.*, 65, 404–418.

Jones, M., Buchana, G., Harper, R. and Xech, P-L. (2007) Questions not answers: a novel mobile search technique. In *CHI 2007*, 155–158.

Kaikkonen, A., Kallio, T., Kekáláinen, A., Kankainen, A. and Cankar, M. (2005) Usability testing of mobile applications: a comparison between laboratory and field testing. *J. Usability Stud.*, 1, 4–16.

Kjeldskov J. and Graham C. (2003) A review of mobile HCI research methods. In *Proceedings of the 5th International conference on Mobile HCI, Mobile HCI 2003*, Udine, Italy. LNCS, Springer-Verlag, 317–335.

Kjeldskov, J. and Stage, J. (2004) New techniques for usability evaluation of mobile systems. *Int. J. Human-Comp. Stud.* 60, 599–620.

Kline, P., (1986), A Handbook of Test Construction. Methuen.

Krueger, R. A. and Casey, M. A. (2000), *Focus Groups: A Practical Guide for Applied Research* (3rd edn), Sage.

D. E. Knuth. (1988) *The Art of Computer Programming*. Addison-Wesley, Reading, MA.

Larsen, L. B. (2003a) Assessment of spoken dialogue system usability? what are we really measuring? In *Proceedings of 8th European Conference on Speech Communication and Technology, Eurospeech 2003*, Geneva, Switzerland. ISCA, 1945–1948.

Larsen, L. B. (ed.) (2003b) Evaluation methodologies for spoken and multi modal dialogue systems. COST278 WG2 and WG3 Report.

Lewis C. and Rieman, C. (1993). Task-Centered User Interface Design: A Practical Introduction. http://hcibib.org/tcuid/.

Kaj Mákelá, Belt, S., Graanblatt, D., Hákkilá, J. (2007) Mobile interaction with visual and RFID tags – a field study on user perceptions. In *CHI 2007*, 991–994.

Melto A., Turunen M., Hakulinen J., Kainulainen A. and Heimonen T. A (2008) Comparison of input entry rates in a multimodal mobile application, In *Proceedings of Interspeech 2008*.

Móller, S. (2002) A new taxonomy for the quality of telephone services based on spoken dialogue systems. In *Proceedings of the 3rd SIGdial Workshop on Discourse and Dialogue* Philadelphia, PA, USA. ACL.

Möller, S. (2005) Parameters for quantifying the interaction with spoken dialogue telephone services. In *Proceedings of the 6th SIGdial Workshop on Discourse and Dialogue*, Lisbon, Portugal. ACL, 166–177.

Möller, S. and Skowronek, J. (2003) Quantifying the impact of system characteristics on perceived quality dimensions of a spoken dialogue service. In *Proceedings of the 8th European Conference on Speech Communication and Technology, Eurospeech 2003*, Geneva, Switzerland. ISCA, 1953–1956.

Nielsen, J. (1994). *Usability Engineering*. Morgan Kaufmann.

O'Hara, K., Slayden Mitchell, A. and Vorbay, A. (2007) Consuming video on mobile devices. In *CHI 2007*, 857–866.

Palen, L. and Salzman, M. (2002) Voice-mail diary studies for naturalistic data capture under mobile conditions. In *Proceedings of CSCW02*, New Orleans, Louisiana, USA, 87–95.

Parasuraman, A., Zeithaml, V. A. and Berry, L. L. (1988) SERVQUAL: A multiple-item scale for measuring consumer perceptions of service quality. *J. Retailing*, 64, 12–40.

Po, S., Howard, S., Vetere, F. and Skov, M. B. (2004) Heuristic evaluation and mobile usability: bridging the realism gap. In *MobileCHI 2004*, LNCS 3160, 49–60.

Rajput, N., Nanavati, A. A., Kumar, M., Kankar, P. and Dahiya, R., SAMVAAD: speech applications made viable for access-anywhere devices. In *IEEE International Conference on Wireless and Mobile Computing, Networking and Communications (WiMob2005)*, Montreal, Canada, August 2005.

Salonen E., Turunen M., Hakulinen J., Helin L., Prusi P. and Kainulainen A. (2005) Distributed dialogue management for smart terminal devices, In *Proceedings of Interspeech 2005*, 849–852.

Salovaara, A. (2007) Appropriation of a MMS-based comic creator: from system functionalities to resources for action. In *CHI 2007*, 1117–1126.

Seager, W. and Stanton Fraser, D. (2007) Comparing physical, automatic and manual map rotation for pedestrian navigation. In *CHI 2007*, 767–776.

Shneiderman, B. The limits of speech recognition, *Comm. ACM*, 43, 63–65.

Smeele, P. M. T. and Waals, J. A. J. S. (2003) Evaluation of a speech-driven telephone information service using the PARADISE framework: a closer look at subjective measures. In *Proceedings of 8th European Conference on Speech Communication and Technology, Eurospeech 2003*, Geneva, Switzerland. ISCA, 1949–1952.

Turunen M., Hakulinen J., Salonen E., Kainulainen A. and Helin L. (2005) Spoken and multimodal bus timetable systems: design, development and evaluation. In *Proceedings of 10th International Conference on Speech and Computer (SPECOM 2005)*: 389–392.

Turunen M., Hakulinen J. and Kainulainen A. (2006) Evaluation of a spoken dialogue system with usability tests and long-term pilot studies: similarities and differences. In *Proceedings of Interspeech 2006*. ICSLP, 1057–1060.

Turunen M., Hakulinen J., Kainulainen A., Melto A. and Hurtig T. (2007) Design of a rich multimodal interface for mobile spoken route Guidance. In *Proceedings of Interspeech 2007 - Eurospeech*, 2193–2196.

Turunen M., Melto A., Hakulinen J., Kainulainen A. and Heimonen T. (2008) User expectations, user experiences and objective metrics in a multimodal mobile application, In *Proceedings of SiMPE 2008*.

Walker, M. A., Litman, D. J., Kamm, C. A. and Abella, A. (1997) PARADISE: A framework for evaluating spoken dialogue agents. In *Proceedings of the 35th Annual Meeting of the Association for Computational Linguistics*, Madrid, Spain. ACL, 271–280.

Walker, M. A., Litman, D. J., Kamm, C. A. and Abella, A. (1998) Evaluating spoken dialogue agents with PARADISE: Two case studies. *Comp. Speech Lang.*, 12, 317–347.

Walker, M. A., Kamm, C. A. and Boland, J. (2000) Developing and testing general models of spoken dialogue system performance. In *Proceedings of the 2nd International Conference on Language Resources and Evaluation, LREC 2000*, Athens, Greece.

Walker, M. A., Passonneau, R. and Boland, J. E. (2001) Quantitative and qualitative evaluation of DARPA communicator spoken dialogue systems. In *Proceedings of the 39th Meeting of the Association of Computational Linguistics, ACL 2001*, Toulouse, France. ACL, 515–522.

Weiss, S. (2002) *Handheld Usability*. John Wiley & Sons.

Wharton, C., Rieman, J., Lewis, C. and Polson, P. (1994). The cognitive walkthrough method: a practitioner's guide. In J. Nielsen and R. Mack (eds) *Usability Inspection Methods*, 105–140, John Wiley & Sons.

White, S., Marino, D. and Feiner, S. (2007) Designing a mobile user interface for automated species identification. In *CHI 2007*, 291–294.

9

Developing regions

Nitendra Rajput and Amit A. Nanavati
IBM Research, India

More than 80% of people living on Earth do not access the internet. Reasons include lack of affordability, internet penetration, literacy and relevance. However, in places where these people live, phone penetration, and particularly mobile phone penetration, is high. More and more of these people are getting access to the phone, and many of them cannot read or write. This makes the mobile phone the primary platform, and speech the primary interface for these people.

This makes a compelling case for supporting speech applications for mobile phones in local languages. While the mobile phones used by the 'bottom of the pyramid" may not yet be feature-rich and able to support much client-side processing, the screen and the touchpad offer multimodal support for voice user interfaces.

So, we are now faced with a great opportunity to build speech applications in local languages for the mobile. How do we go about building such applications rapidly? How do we build automatic speech recognizers (ASRs) for these local languages? How to build intuitive and compelling voice user interfaces for these applications?

While we have been developing speech applications for a while now, and are familiar with the experiences and frustrations of dealing with speech-based systems, we need to keep in mind that this is a completely new target segment.

Speech in Mobile and Pervasive Environments, First Edition.
Nitendra Rajput and Amit A. Nanavati.

We will have to start afresh to understand this new set of users: what applications they find relevant and what types of interaction do they prefer? Given the diversity of the user base, we may well find that what works in Africa may be different from what works in India.

9.1 Introduction

In a way, this chapter could have been a miniature version of this entire book, with a developing region slant. Given the special needs of this new user base, it is not surprising that hardware, software, context and user-interfaces will all have to be appropriately designed to meet them. The hardware has to be cost-effective and robust, and along with the software needs to support the most convenient forms of interaction. In some cases, noise will be a bigger barrier to cross. Leveraging the cultural and local context to mitigating the effect of noise and supporting ASR could become important. Perhaps the kind of applications and the ways of accessing them will be different, and so will the means of evaluating them.

The reason that this chapter is not a miniature version of this book is that this is an area of recent activity, and most of these questions are still unanswered. Therefore, we shall attempt to assemble the handful of multi-disciplinary efforts that are ongoing.

9.2 Applications and studies

While speech appears to be an attractive option in principle, practical considerations such as dialects, accents and lack of robust ASRs for local languages can be potential deterrents.

Plauché *et al.* (2006b) describe an inexpensive approach to gathering the linguistic resources needed to power a simple spoken dialog system. Data collection is integrated into dialog design: users of a given village were recorded during interactions, and their speech semi-automatically integrated into the acoustic models for that village, thus generating the linguistic resources needed for automatic recognition of their speech. Plauché and Prabaker (2006) conducted a pilot user study for Tamil Market, a speech-driven agricultural query system, in community centers in rural India. They report that preliminary findings from a Wizard-of-Oz field study showed that rural villagers are able to navigate through a dialog system using their voice, regardless of literacy levels and previous experience with technology, while traditional user study techniques favored literate users.

Another set of studies by Sherwani *et al.* (2009) and Patel *et al.* (2006) compared the efficacy of speech versus dual-tone multiple frequency (DTMF) (key pressing) as input modalities. Sherwani *et al.* (2009) found that well-designed speech interfaces significantly outdid touchtone equivalents for both low-literate and literate users, and that literacy significantly impacts task success for both modalities. On the other hand, Patel *et al.* (2006) found that the task completion rates were significantly higher with dialed input, particularly for subjects under age 30 and those with less than an eighth-grade education. Additionally, participants using dialed input demonstrated a significantly greater performance improvement from the first to final task, and reported less difficulty providing input to the system.

A similar study by Grover *et al.* (2008), in the context of providing health information to caregivers of HIV-positive children, was performed in Botswana with semi- and low-literate users. Their results indicate a user preference for touchtone over speech input, although both systems were comparable in performance based on objective metrics. Whilst this pilot study illustrated that telephony services could in fact be easily used by semi- and low-literacy users, and that a spoken dialog system in a local language can be a powerful health education tool, the decisive factor in widespread uptake is likely to be the cost incurred by the caller for the service. The majority of caregivers said that even though the service would be useful to them they would only be able to make use of it if the service were toll-free.

9.2.1 VoiKiosk

Agarwal *et al.* (2009) experimented with a voice-based kiosk solution, VoiKiosk, for people in rural areas. This system is accessible by phone and thus meets the affordability and low-literacy requirements. The authors describe usability results gathered from usage by more than 6500 villagers during the eight months of the system was deployed in the field. Their experiments suggest the importance of locally created content in their own language for this population. The system provides interesting insights about the manner in which this community can create and manage information. Based on the use of the system in the eight months, VoiKiosk also suggests a mechanism to enable social networking for the rural population.

A VoiKiosk act as information and service portal for a village. It can be a central point of access for a community where information relevant to the community can be posted and accessed directly by the users themselves. This solution does not rely on internet connectivity, which is most often not available in rural areas. Most importantly, it allows end-users to directly interact with the services, thus removing any dependence on a kiosk operator. The following are the four main categories of information available on the deployed VoiKiosk system for the village:

- *V-Agri*. Farmers use this service to consult agricultural experts regarding their crop-related problems. Prior to the advent of VoiKiosk, a picture of the crop was taken and sent to an expert, who then sent a reply back to the farmer through the local government. The turn-around time for this process was 24 hours. With VoiKiosk, the expert is able to post his advice for the farmer on the VoiKiosk, reducing the turn-around time to 4 hours.

- *Health information*. Information related to different health advisory and health camps is posted in the VoiKiosk. The schedule of the doctor's visit to the health center is also posted on the VoiKiosk. The VoiKiosk administrator can change the message if there is a change in the timing of the doctor's visit.

- *Distance education schedule*. Information regarding new programs and a schedule of daily classes or changes in the schedule are posted to the VoiKiosk.

- *Professional services*. In this section of the VoiKiosk users (including mechanics, drivers and truck-owners) are able to place personal advertisements.

The VoiKiosk system was live, 24 hours a day and seven days per week for eight months. Over this period, the system received a total of 114 782 calls from 6509 villagers. Usage patterns show that one or both of the following must be true of the users:

1. This population is more patient and can listen to a longer list of information over phone than western populations have been observed to do with spoken dialog systems.

2. The villagers see a certain value in posting an advertisement and so are ready to wait a long time in order to get a chance to post.

The increasing use of the VoiKiosk system for different purposes seems to demonstrate that a voice-based mechanism for local content creation is a very powerful interaction modality for the provision of information and communication technologies in rural areas.

Sherwani *et al.* (2007) have developed VoicePedia, a telephone-based dialog system to enable purely voice-based access to the unstructured information on websites such as Wikipedia. The VoicePedia interface was designed to mimic the web search experience, but is site-constrained to the Wikipedia domain. The interaction consists of three major dialog phases: (a) keyword entry, (b) search result navigation and (c) web-page navigation. In the keyword-entry phase the user is asked to say search keywords one at a time. After each keyword, the system uses one of two disambiguation strategies – Best Guess or Web Suggest – to choose a hypothesis from the n-best list, (which is a list of top in possible results, one of which is expected to contain the right result), and then reports this hypothesis to the user, who may then either correct it, or continue with the next keyword. Once all keywords have been entered, the user says 'that's all' to move to the next phase. Based on the search keywords, the system performs a web search query of the form: `site:en.wikipedia.org` k_1 k_2 k_3 k_n where k_i represents the ith keyword. The user is then presented with a list of the page titles of the top 10 search results, with pauses after the 3rd and 6th results. The user is able to select a search result by repeating the title or the search result number, or by interrupting the replay of the list after hearing a specific search result, or by asking for the first or last result. Users found that keyword entry through the voice interface was significantly faster than search result navigation and page browsing. Although users preferred the GUI-based interface, task success rates for the two systems were comparable.

9.2.2 HealthLine

Sherwani *et al.* (2007) have built a system for speech-based health information access . Health information access by low-literate community health workers is a pressing need in community health programs across the developing world. The authors conducted a needs assessment to understand the health information access practices and needs of various types of health workers in Pakistan. They used Microsoft Speech Server 2007 Beta (MSS), which includes a speech recognition engine, a speech synthesis engine and a dialog management architecture, along with dialog authoring tools. However, MSS supports only a few languages, not including Urdu. To solve this problem for speech synthesis, all audio was recorded as individual prompts from an Urdu-speaking voice talent. It was then possible to encode prompt text in Urdu in the code (through Unicode), so that there was no extra layer of mapping required. Most (if not all) speech recognition systems allow the definition of new words in a lexicon file, this file containing each word's textual representation and its phonetic representation(s). The authors defined each of the Urdu words that needed to be recognized by the system, and hand-coded pronunciation using US English phonemes. Once all the words were defined this way, context-free GRXML grammars could be created to recognize utterances.

The study was conducted at the Basic Health Unit in Wahi Pandi, in Dadu District, Sindh, Pakistan. Environmental noise was a frequent problem, and one method that was tried to deal with this showed promise: by lowering the input volume on the microphone, the system would only hear the loudest of sounds. Many times, this would mean that it would not even hear the user speak, but the users quickly learned to speak louder so as to avoid having to hear the prompt again. Of the six participants who used the speech interface, five were able to successfully hear and report the information they heard. Most significantly, the low-literate participant who was not able to read pamphlets at all was able to use the speech interface successfully.

9.2.3 The spoken web

In Kumar *et al.* (2007a), the authors describe the spoken web, also known as the World Wide Telecom Web (WWTW or sometimes T-Web) – their vision of a voice-driven ecosystem parallel to that of the World Wide Web. WWTW is a network of interconnected *voicesites*, which are voice driven applications created by users and hosted in the network. Specifically, WWTW

- enables the underprivileged to create, host and share information and services produced by themselves;
- provides simple and affordable access mechanisms to let ordinary people exploit IT services and applications that are currently only available to web users; and
- provides a cost-effective ecosystem that enables users to create and sustain a community parallel to the World Wide Web.

WWTW voicesites are identified by global identifiers called *voinumbers* and can be interconnected through *voilinks*. A voinumber is a virtual phone number that either maps onto a physical phone number or other uniform resource identifiers such as a SIP URI. A voicesite is a voice-driven application that consists of one or more voice pages (e.g. VoiceXML files) that are hosted on the telecom infrastructure. Voicesites are accessed by calling up the associated voinumber and interacting with its underlying application flow through a telephony interface. A voilink is a link from one voicesite to another through which a caller interacting with the source voicesite can be transferred to the target voicesite in the context of the voicesite application. The WWTW can therefore be visualized as a system that operates over the telecom infrastructure and parallels can be drawn with the World Wide Web that runs on internet infrastructure.

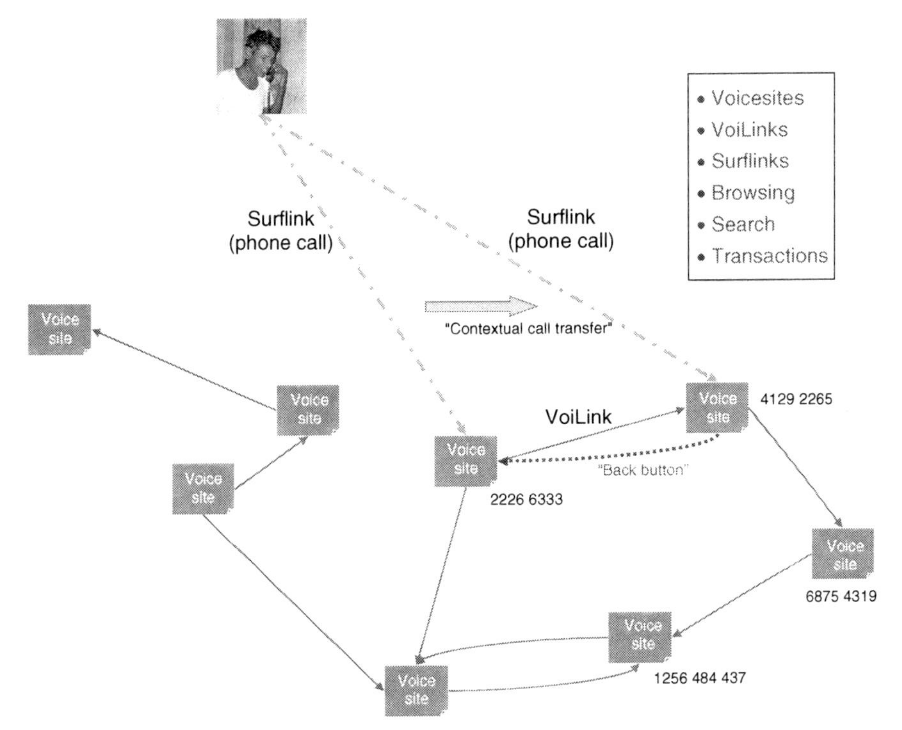

Figure 9.1 The World Wide Telecom (Spoken) Web of voicesites, connected by Voilinks, can support browsing and transactions, all by talking over the phone.

The WWTW model can be represented as a graph $G = (V, E)$ where $v \in V$ is a voicesite and $e \in E$ is defined as an edge from v_1 to v_2 if there exists a VoiLink from v_1 to v_2, where $v_1, v_2 \in V$.

Voicesites, voinumbers and voilinks form the basic building blocks of WWTW. The VoiGen system, described in Section 9.3, is a template-based voicesite authoring system. It is analogous to HTML page-authoring software for websites.

HSTP

HTTP provides a mechanism to connect web sites. The success of the World Wide Web can be partly attributed to the seamlessly browsable web that is formed through this connectivity. In Agarwal *et al.* (2008), the authors define hyperspeech as a voice fragment in a voice application that is a hyperlink to a voice fragment in another voice application. They describe *Hyperspeech Transfer Protocol* (HSTP), a protocol to seamlessly

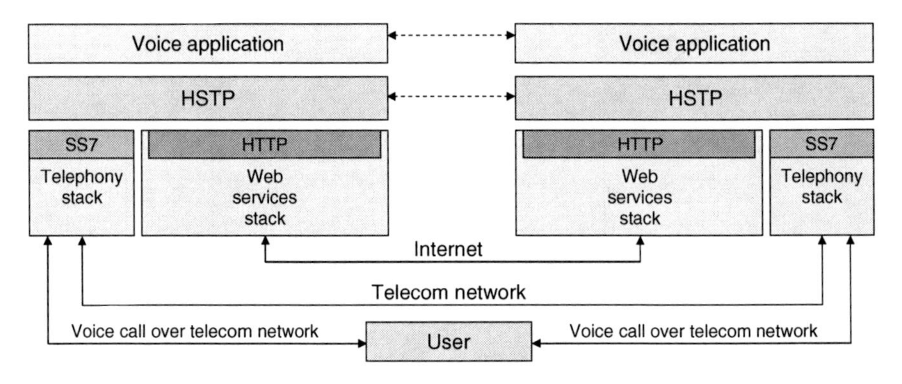

Figure 9.2 The HSTP stack (Agarwal et al., 2007).

connect telephony voice applications. HSTP enables users to browse across voice applications by navigating the hyperspeech content in a voice application. HSTP can also be used for developing cross-enterprise applications that allow a user to transact across two or more voice applications. Apart from enabling secure cross-enterprise transactions, HSTP also allows navigation across voice applications, potentially hosted by different enterprises. It enables the concept of links between voice applications and provides the user with the ability to browse forward and backward across voice applications.

Spoken web browser

In Agarwal *et al.* (2008), the authors describe the WWTW browser, which enables users to experience the same browsing experience as is available on the web. Since the WWTW can be accessed through a dumb phone instrument, the WWTW browser actually resides on the network (instead of the client). The browser can be accessed by making a phone call to a specific number, and standard browsing features such as back, forward, bookmarking and history are supported. The working of the T-Web browser can be seen using the circled steps shown in Figure 9.3:

1. The user calls the T-Web browser to access a voicesite.
2. The T-Web browser transfers the call to the phone number of the voicesite through HSTP.
3. When a user selects a hyperspeech link to browse to the other voicesite, the session is transferred to the target voicesite and HSTP passes the call transfer information to the T-Web browser.
4. This information is stored in the browser history.

5. The user issues a browser command, for example 'go back'.
6. The T-Web browser instructs the HSTP layer on the current voicesite to initiate a transfer to the earlier voicesite phone number.
7. At any time, the user can say 'bookmark' to bookmark the currently browsed voicesite.

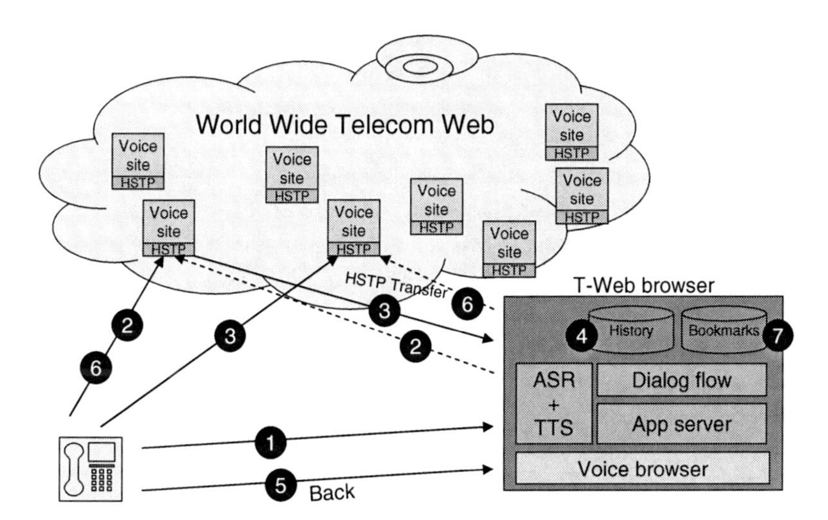

Figure 9.3 The T-Web browser interactions with voicesites.

9.2.4 TapBack

Robinson *et al.* (2011) created and studied 'TapBack', a back-of-device tap to control a dialled-up, telephone-network-based voice service.

Although both ASR and DTMF allow a level of control and interaction with audio content, Robinson *et al.* (2011) believe there is still much work to be done in terms of improving the expressiveness and range of interactions. As a first step toward richer mobile voice interfaces, the authors developed TapBack, an extended interaction method for voice sites that aims to allow callers to smoothly navigate through and control the content they are listening to without having to unnecessarily interrupt its playback. Their approach uses simple back-of-device interactions –audio gestures–on the phones users already own.

The TapBack system allows callers a richer experience with interactive voice sites by enabling audio gestures to be used at any time during a call. By using the back of user's phones as an input surface while a call is in progress, the interruptions of ASR/DTMF are removed and this allows users to keep the phone by their ear throughout the call. Unlike previous back-of-device methods, the phone's inbuilt microphone is used to pick up sounds generated on the back of its case. These sounds are loud enough to be transferred to the other party in the call but, unlike DTMF tones, are not so loud at the caller's end that they drown out the audio being played.

For a simple user introduction to audio gestures we chose to apply tapping recognition to the control of audio playback speed. Previous voice-site analyzes (Dhanesha *et al.* 2010) have shown that callers appreciated finer control of playback, so this was a natural application for our system. In our implementation, when users tap two or three times, playback rates can be increased by 25% or 35%, respectively, speeds which maintain the intelligibility of the audio. Tapping once returns playback to its normal speed.

The TapBack system is installed on a remote server, which monitors low-level network packets to track incoming phone calls to individual voice sites. When a call is established, real-time audio capturing, decoding and analysis is initialized. The audio is first filtered to remove frequencies below 3 kHz, greatly reducing the problem of ambient noise. The stream is then windowed using a 512-sample Hamming window with an overlap of 7/8. Tap recognition itself is very unsophisticated, simply searching each window for short, high-intensity, high-frequency sounds. Detected tap events are then fed to a higher-level audio gesture classifier, which uses timeouts and basic heuristics to classify each tap type. When an audio gesture is found, the system sends a command to the SpokenWeb server, which adjusts the voice-site playback speed in response to the request. Users are also free to control the speed of playback by using DTMF inputs instead of taps. In this case, keys 4, 5 and 6 correspond to single, double and triple taps, respectively.

The TapBack system was evaluated in three ways, focusing primarily on the viability of the approach rather than the sophistication of the recognizer. Recognition accuracy was measured and refined using a test voice site. A live deployment of the system over an extended period was used to assess the usefulness and usability of the approach. Finally, we explored alternative audio gestures with participants to understand the potential for extending such techniques.

The authors conducted a user study to measure and improve the recognizer's accuracy over a standard telephone connection. Eighteen users of an existing, popular farming information voice site based in a rural region in India were recruited (see Patel *et al.* (2010) for more detailed user population demographics). To ensure a cross-section of user expertise, participants included people who accessed the voice site very regularly and also those who were only casual users. All users were male, and the average age was 32. The set of phones used by the participants consisted of 14 different (low-end) handset types produced by four manufacturers. All participants had already used the

DTMF speed-control methods detailed in Dhanesha *et al.* (2010). Each participant was called by phone to explain the study method and the concept of audio gestures. The calls were made to participants when they were at the locations from which they usually interacted with the voice-site services. The participant was then connected to a test voice site, which asked them to tap the back of the phone while holding it to their ear, in response to four sets of cues. Each cue set asked users to tap once, twice, then three times. Each participant therefore provided 12 tap commands. Recognition rates were: 1 tap, 93%; 2 taps, 78%; 3 taps, 56%. High accuracy rates for single and double taps were encouraging given:

- the minimal explanation of the concept to users; this form of interaction with a service was entirely novel in users' experience;
- the diverse set of low-end phones involved;
- the audio channel being of standard telephone quality;
- the study being in a live setting and not in a laboratory.

A large proportion of the errors in recognition were due to participants tapping more slowly than the recognizer expected. This led to two taps being recognized as 1+1 taps (accounting for 50% of the two-tap errors) and three taps being recognized as 1+2, 2+1 or 1+1+1 taps (60% of three-tap errors). To deal with these errors, the tap classifier was modified to employ simple correction heuristics so that, for example, a 2-tap shortly followed by a 1-tap was interpreted as a 3-tap instruction. The remaining errors were caused by taps not being distinct enough for the recognizer to extract it from the input.

The TapBack system was made available on a live farming information voice site (Patel *et al.* 2010). This exploratory study aimed to measure the adoption of audio gestures by logging any tap interactions and responses during normal use of the service. The system was deployed for 12 days, during which any of the 110 registered active users could call at any time. These users were geographically dispersed over a wide area of India, and were all farmers living in rural settings. When calling, users were given a brief automated introduction to the new method, which explained how they could tap the back of their phone to control the playback speed. The system logged call details and any input actions (both tap-based and DTMF). This data was supplemented by conducting detailed telephone interviews with 15 users. Ten of these were selected at random from the set of those who had used TapBack during the deployment period; the remaining five were randomly selected from callers who did not attempt to use the tap interaction. The average age of participants was 31, and all except one were male. During the interviews these users were asked about their reactions to the approach: how usable it was and any issues they had encountered in its use. Interviews were conducted in Gujarati, the participants' native language.

A total of 286 calls to the voice site were recorded over the study period, from 52 unique callers. 1293 tap interactions were recorded in total and 36 callers used the TapBack feature (166 calls; 7.8 taps per call on average). Of the 36 participants that used tap interaction, 25 used the feature on more than one call. Two others called more than once but only used tap interaction on their first call; the remaining 9 TapBack users called only once over the study period. The 16 participants who did not use tap interactions did not use DTMF for speed control, either. Tap interactions consisted of 772 single, 301 double, and 220 triple taps. Few of the callers that wanted to control the speed of the call used the DTMF method. 52 speed-control DTMF events were recorded in total.

Of the the ten callers who had used the TapBack feature, the majority were positive in their comments about the approach. Benefits mentioned ranged from those related to utility to those concerning the less-tangible user experiences. Several respondents commented on the tapping being easier to use and quicker than DTMF. Another interviewee talked of the fun of the new interaction. Interestingly, one participant said, that is was like having a touchscreen, that it was a modern thing to use, and that is was 'cool'. Some of the negative comments from these ten adopters were predictable, such as expressing frustration when a tap-event was not recognized: one respondent said he would always use the buttons because they always worked–'end of story'. However, there were also issues related to the use context. Two interviewees worried about using the system regularly as the tapping, to their mind, might damage the phone. For one of these interviewees this was particularly worrying as they often lent their phone to others to use (a practice quite common in rural areas). Another respondent said they tended to listen to the service with a group of people using the speakerphone.

There were two explanations for the non-use of the approach by the five other interviewees. For some the environment (as witnessed during telephone interviews) was too noisy, while others had not understood the explanation of the new feature given by the voice site after call connection.

The logged data provides evidence that callers are willing to adopt the tapping method, with the majority of callers using the approach. It should also be noted that the functions controlled by TapBack – speeding and slowing the audio – are optional: users are able to listen to and navigate through content without employing them. We would expect, then, that some calls during the study would not show any tap interactions. Moreover, 93% of callers who used TapBack during their first call also used it again in their subsequent calls.

Callers that used TapBack did so several times in each call. The comments about the system's utility value are, of course, encouraging, especially when considering the accuracy of our simple recognizer. However, of more note, perhaps, are the responses relating to the user experience–a fun, modern interaction is something that is not usually associated with the low-end devices these users have access to. The negative comments are spurs to improve the recognition engine and explanation of its use. The social issues raised suggest extensions to our approach: we will need to ensure that the tap-models used are tuned not just to individual phone numbers but to the set of users that might use that phone. In speaker-mode use, we might be able to consider a wider set of audio gestures as suggested in Harrison (2008).

9.3 Systems

ASR is the process of algorithmically converting a speech signal (audio) into a sequence of words (text). This is usually achieved by training hidden Markov models (HMMs) for phones, diphones or triphones from training data, a hand-labeled speech corpus.

Adaptation refers to the process of tuning the speech recognizer using target speech data obtained from the field to better match the training and test conditions. The problem therefore is to adapt an ASR system from a source language, for which training resources (transcribed speech data) are available, to a target language for which there are limited or no annotated corpora. Plauché *et al.* (2006a) built a Tamil language speech recognizer trained on a database of speech recordings of 80 speakers. The speech recognizer used triphone HMM models (single Gaussian) and state-based parameter-tying for robust estimation. Decision-tree-based, state-tied, triphone models easily accommodate new words and contexts by traversing through the tree and synthesizing the triphones from a cluster of acoustically similar models. The test database was created by recording the 28 vocabulary options three times each. Monophone models yielded 73% accuracy while triphone models performed at 97%.

Waibel *et al.* (2000) recommend that *cross-language transfer* should be used when no existing data is available for a language. This involves training the recognizer on one or more source languages. Linguistically similar models and multilingual models offer the best results. *Language adaptation* should be used to generate linguistic resources for a target language by initializing acoustic models on available source language data and adapting the models to the target language using a very limited amount of training data. Performance correlates to the amount of data available in the target language and the number of different speakers used for training is found to increase performance more than the number of utterances.

A strategy that takes advantage of the substantial overlap of speech sounds across languages is to train acoustic models on the IPA (International Phonetic Alphabet) representation (Schultz and Waibel 1998). Training acoustic models directly on graphemes (script characters) or automatically converting graphemes to phonemes omits the need to create a pronunciation dictionary by hand (Kanthan and Ney 2003). This method works best for languages with a close grapheme-to-phoneme relationship.

Another set of approaches, used for many years, has been to create custom hardware for speech recognition (Hon 1991). Recently, Nedevschi *et al.* (2005) argue for revisiting this approach because:

- very low-power solutions are required for successful hand-held use in developing regions, and custom hardware can be many times more power-efficient than software-based solutions;

- as opposed to the past, today's abundance of embedded devices could make the use of speech-based interfaces ubiquitous;

- memory bandwidth limitations, a traditional bottleneck in parallelizing speech computation, can be easily overcome using today's technology by integrating multiple blocks of FLASH or SRAM memory along with logic on the same chip.

The authors propose a system architecture for real-time hardware speech recognition on low-cost, power-constrained devices. The system is intended to support real-time speech-based user interfaces for the developing world. Their system is built on top of a physical infrastructure composed of a set of low-cost handheld devices, featuring speech-based user interfaces and one or more central servers. The expensive and lengthy training operation is performed on shared servers. Since permanent network connectivity with the servers cannot be assumed, the handheld devices support the entire recognition chain of speech recording, spectral analysis and decoding. The spectral analyzer front-end requires digital signal processing primitives, such as fast Fourier transforms, that can be efficiently supported in custom hardware. The handheld devices periodically record speech samples along with the transcriptions. The samples are transferred to the server, where they retrain the speech models for higher accuracy, and possibly in a way that is speaker dependent. These new models are then loaded onto the devices at the next device-server communication opportunity. The user interface communicates with the the recognition engine through a dedicated driver. The application supplies the recognizer with context information, specifying the set of allowed word combinations in the given context, and the recognition engine responds with the most probable set of words.

Kumar *et al.* (2007b) created VoiServ system, which enables ordinary telephone subscribers to create, deploy and offer their own customized voice-driven applications called voicesites. The generated voicesites get hosted in the network for low cost of ownership and maintenance, and are integrated with advanced services available in the converged networks of today. Figure 9.4 depicts the architecture of VoiServ system consisting of VoiGen, a voice driven generator of voice based applications, along with VoiHost, a voicesite hosting engine. VoiGen simplifies the process of creation of voice-based applications by enabling it through a voice-driven interaction. A phone subscriber could call in to VoiGen and compose an application by navigating through the custom options offered to her. This application is then deployed in the form of a voicesite. VoiGen makes use of existing components (reusable dialogs as well as IT components such as databases and web services) to compose custom applications. By virtue of having a voice-driven interface, the services get exposed to all telephony devices.

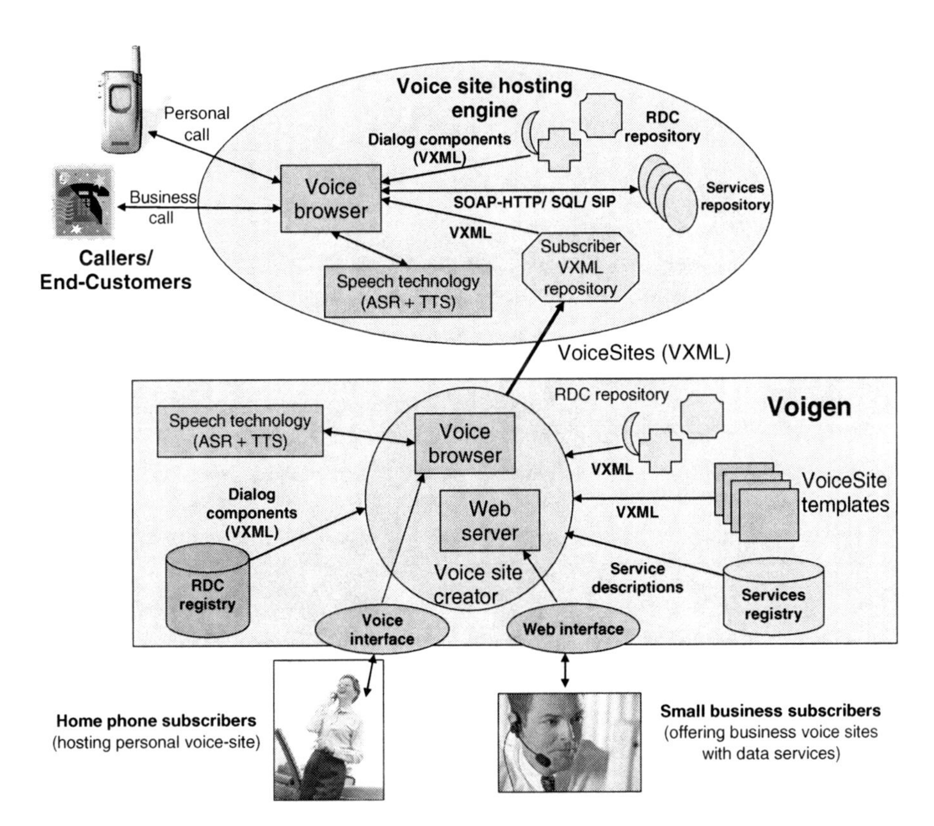

Figure 9.4 Voigen system architecture.

9.4 Challenges

This is a very new area where a lot of activity is taking place currently. There are naturally many interesting questions to be answered still. Here we highlight just a few of the problems that we feel are important to address if speech is to be effective as a medium, enabling mobiles to become the main platform for communication in developing regions:

- speech recognition in noisy environments;
- novel and easier ways of handling multiple languages and dialects;
- speech systems with small footprints of memory and power consumption;
- leveraging context (such as location and profile) to make more intelligent voice user interfaces that reduce the cognitive burden of semi-literate and non-literate users;
- effective navigation of speech applications;
- special hardware for on-device recognition support;
- exploiting the multimodality of mobile phones.

Bibliography

Agarwal, S. K., Kumar, A., Nanavati, A. A and Rajput, N. (2009) Content creation and dissemination by-and-for users in rural areas. *ICTD 2009*, Doha, Qatar.

Agarwal, S. K., Jain, A., Kumar, A. and Rajput, N., (2010) The World Wide Telecom Web browser. *SIGDEV 2010*, London, UK.

Agarwal, S. K., Chakraborty, D., Kumar, A., Nanavati, A. A. and Rajput, N. (2007) HSTP:The hyperspeech transfer protocol. *ACM Hypertext and HyperMedia*.

Dhanesha, K. A., Rajput, N. and Srivastava, K. (2010) User driven audio content navigation for spoken web. In *Proceedings Multimedia 10*, 1071–1074.

Franz, A. and Milch, B. (2002) Searching the web by voice. P*roceedings of the 19th International Conference on Computational Linguistics (COLING)*. 1213–1217.

Grover, A. S., Plauché, M., Barnard, E. and Kuun, C. (2008) HIV Health information access using spoken dialog systems: touchtone vs. speech. (Manuscript) CSIR, Meraka Institute, South Africa.

Harrison, C. and Hudson, S. E.. (2008) Scratch input: creating large, inexpensive, unpowered and mobile finger input surfaces. In *Proceedings of UIST 08*, ACM, 205–208.

Hon, H.-W. (1991) A survey of hardware architectures designed for speech recognition. Technical Report CMU-CS-91-169.

Hussain, F. and Tongia, R. (2007) Community radio for development in south asia:a sustainability study. *Proceedings of the IEEE/ACM ICTD 2007*, Bangalore, India.

Kanthak, S. and Ney, H. (2003) Multilingual acoustic modeling using graphemes. In *Proceedings of the European Conference on Speech Communication and Technology*, Switzerland.

Kumar, A., Rajput, N., Chakraborty, D., Agarwal, S. K. and Nanavati, A. A. (2007a) WWTW: The World Wide Telecom Web. *SIGCOMM NSDR,* ACM.

Kumar, A., Rajput, N., Chakraborty, D., Agarwal, S. K and Nanavati, A. A. (2007b) VoiServ: creation and delivery of converged services through voice for emerging economies. World of wireless, Mobile and Multimedia Networks, Espoo, Finland, 1–7.

Nedevschi, S., Patra, R. K. and Brewer, E. A. (2005) Hardware speech recognition for user interfaces in low cost, low power devices. *DAC 2005*, Anaheim, California, USA.

Patel, N., Chittamuru, D., Jain, A., Dave, P. and Parikh, T. S. (2010) Avaaj Otalo: a field study of an interactive voice forum for small farmers in rural india. In *Proceedings of CHI 10*. ACM, 733–742.

Patel, N., Agarwal, S., Rajput, N., Nanavati, A., Dave, P. and Parikh, T. S. (2006) Comparative study of speech and dialed input voice interfaces in rural India. *CHI 2009*.

Plauché, M. and Prabaker, M., (2006) Tamil Market: A spoken dialog system for rural India. CHI 2006, April 2227, 2006, Montral, Qubec, Canada.

Plauché, M., Cetin, Ö. and Nallasamy, U. (2006a) How to build a spoken dialog system with limited (or no) language resources. AI in ICT4D, ICFAI University Press, India.

Plauché, M., Nallasamy, U., Pal, J., Wooters, C. and Ramachandran, D. (2006b) speech recognition for illiterate access to information and technology. *Proceedings of ICTD 2006*, Berkeley, May 2006.

Ramachandran, D. and Canny, J. (2008) Applying persuasive technologies in developing regions. HCI for Community and International Development, *Workshop at CHI 2008*, Florence, Italy.

Randolph, M., Engelsma, J., Ferrans, J. and Pearce, D. (2007) Architecting mobile and pervasive multimodal applications for developing countries. *2nd SiMPE Workshop, MobileHCI 2007*, Singapore.

Robinson, S., Rajput, N., Jones, M., Jain, A., Sahay, S. and Nanavati, A. A. (2011) TapBack: towards richer mobile interfaces in impoverished contexts. In *Proceedings of of ACM CHI*.

Schultz, T. and Waibel, A. (1998) Multilingual and crosslingual speech recognition. In *Proceedings of DARPA workshop on Broadcast News Transcription and Understanding*, 259–262.

Sherwani, J., Ali, N., Mirza, S., Fatma, A., Memon, Y., Karim, M., Tongia, R. and Rosenfeld, R. (2007) HealthLine: speech-based access to health information by low-literate users. *Proceedings of the IEEE/ACM ICTD 2007*, Bangalore.

Sherwani, J. and Rosenfeld, R. (2008) The case for speech and language technologies for developing regions. In *Proceedings of Human-computer Interaction for Community and International Development Workshop*, Florence, Italy.

Sherwani, J., Yu, D., Paek, T., Czerwinski, M., Ju, Y.-C. and Acero, A. (2007) Voice-Pedia: towards speech-based access to unstructured information. In *Proceedings of Interspeech*, Antwerp, Belgium.

Sherwani, J., Palijo, S., Mirza, S., Ahmed, T., Ali, N. and Rosenfeld, R. (2009) Speech vs. touch-tone: telephony interfaces for information access by low literate users. In *Proceedings of Information & Communications Technologies and Development*, Doha, Qatar.

Sterling, R., O'Brien, J. and Bennett, J. K. (2007) Advancement through interactive radio. *Proceedings of the IEEE/ACM ICTD 2007*, Bangalore.

Waibel, A., Guetner, P. Mayfield Tomokiyo, L., Schultz, T., and Woszczyna, M. (2000) Multilinguality in speech and spoken language systems. In *Proceedings of of IEEE*, 88, 1297–1313.

Index

Speech in Mobile and Pervasive Environments, First Edition.
Nitendra Rajput and Amit A. Nanavati.

CPSIA information can be obtained at www.ICGtesting.com
Printed in the USA
BVOW061613020212

281967BV00004B/1/P

9 780470 694350